MW00843705

Water Waves and Ship Hydrodynamics

2nd Edition

A.J. Hermans

Water Waves and Ship Hydrodynamics

An Introduction

2nd Edition

A.J. Hermans
Technical University Delft
Delft
The Netherlands
ajh@dds.nl

ISBN 978-94-007-0095-6 e-ISBN 978-94-007-0096-3
DOI 10.1007/978-94-007-0096-3
Springer Dordrecht Heidelberg London New York

Library of Congress Control Number: 2010938437

Cover design: eStudio Calamar S.L.

Printed on acid-free paper

Springer is part of Springer Science+Business Media (www.springer.com)

Preface to the Second Edition

This book is a revision and extension of the book published by R. Timman, A.J. Hermans and G.C. Hsiao based on the lecture notes of courses presented by Timman at the University of Delaware in 1971 and by Hermans at the Technical University of Delft. The main topic of the original text is based on linearised free surface water wave theory. For many years the first edition of the book is used by Aad Hermans as material for a course in ship hydrodynamics presented to Master students in applied mathematics and naval architecture at the Technical University of Delft. Influenced by the progress in the research in water waves and especially in ship hydrodynamics the contents of the course has changed gradually. For instance in offshore engineering the topic like the low-frequency motion of objects moored to a buoy has become an important issue during this period. Therefore an introduction in this field has been added. For didactic reasons the very simple rather abstract problem of the motion of a vertical wall is added. The reason to do so is that most effects that play a role can be treated analytically, while for a general three dimensional object some terms can only be obtained numerically. The use of numerical programs is normal practice in this field, therefor an introduction in the theory of integral equations is presented and some specific problems which may arises, such how to avoid non-physical resonance at the so called irregular frequencies may be avoided. In the first edition a derivation of the structure of the equations of motion in all six degrees of freedom is presented. Because the functions derived there are not easily computed in a practical case, we restrict ourselves to the derivation of the equation of motion in one degree of freedom.

Delft, The Netherlands A.J. Hermans

Preface to the First Edition

In the spring of 1971, Reinier Timman visited the University of Delaware during which time he gave a series of lectures on water waves from which these notes grew. Those of us privileged to be present during that time will never forget the experience. Rein Timman is not easily forgotten.

His seemingly inexhaustible energy completely overwhelmed us. Who could forget the numbing effect of a succession of long wine-filled evenings of lively conversation on literature, politics, education, you name it, followed early next day by the appearance of the apparently totally refreshed red-haired giant eager to discuss mathematical problems with keen insight and remarkable understanding, ready to lecture on fluid dynamics and optimal control theory or a host of other subjects and ready to work into the evening until the cycle repeated. He thought faster, knew more, drank more and slept less than any of the mortals; he literally wore us out. What a rare privilege indeed to have participated in this intellectual orgy. Timman's lively interest in almost everything coupled with his buoyant enthusiasm and infectious optimism epitomised his approach to life, No delicate nibbling at the fringes, he wanted every morsel of every course.

In these times of narrow specialisation, truly renaissance figures are, if not extinct, at least a highly endangered species. But Timman was one of that rare breed. His knowledge in virtually all areas of classical applied mathematics was prodigious. I still marvel that while I was his doctoral student in Delft in the late fifties working on a problem in electromagnetic scattering he had at the same time students working in water waves, cavitation, elasticity, aerodynamics and numerical analysis. He was a boundless source of inspiration to his students in all of these varied fields.

His inattention to detail is legendary but this did not hamper his ability to focus on what was really important in a problem. With a wave of his large hand he would dismiss unimportant errors while concentrating on central ideas, leaving to us the task of setting things right mathematically. This nonchalant attitude toward minus signs and numerical factors was probably deliberate. He wanted people to see the forest, not the trees; to focus on the heart of the problem, not inconsequential superficialities. He had little use for the all too prevalent penchant for examining someone's work looking for errors. He would read a paper looking for the gold, not the dross; looking for what was right, not what was wrong.

Of course this did not make life easy for those around him but it did make it interesting. This will be attested to by George Hsiao and Richard Weinacht whose revised version of the notes from Timman's water wave lectures appeared as a University of Delaware report. Timman and Hsiao then planned to further revise and expand these notes and publish them in book form, but the project came to an abrupt halt with Reinier Timman's untimely death in 1975. It might have remained unfinished had not Aad Hermans' visit of Delaware in 1980 breathed new life into it. Together George Hsiao and Aad Hermans have completed the task of revising the notes, reorganising the presentation, restoring the factors of 2 which Timman had cavalierly omitted, and adding some new material. The first four chapters are based substantially on the original notes, while the fifth chapter and appendices have been added.

It is gratifying to see the completion of these notes. It is not unreasonable to hope that they will provide a useful introduction to water waves for a new generation of mathematicians and engineers. This area was perhaps first among equals in the broad spectrum of Timman's interests. If these notes succeed in stimulating a new generation to concentrate on the challenging problems remaining in this field, they will serve a fitting memorial to a remarkable man whose like will not be soon seen again.

Newark, Delaware
March, 1985

R.E. Kleinman

Acknowledgements to the First Edition

The authors wish to thank Professor Willard E. Baxter for his active interest in the publication of these lecture notes. We are grateful to Professor Richard J. Weinacht who spent a great deal of time helping one of us (GCH) in preparing the original draft of the water wave notes on which the present first four chapters were based. We owe special debt to Professor Ralph E. Kleinman without whom this project probably would never have started.

We would like to express our appreciation to Mr. G. Broere for making most of the drawings, and to Mrs. Angelina de Wit for typing the final manuscript.

Finally, we would like to express our gratitude to the Department of Mathematical Sciences of the University of Delaware, the Department of Mathematics and Informatics of the Delft University of Technology and the Alexander von Humboldt Stiftung for financial support at various stages of the preparation of this manuscript.

March, 1985

A.J. Hermans
G.C. Hsiao

Contents

Chapter 1
Theory of Water Waves

This chapter contains the formulation of boundary and initial boundary value problems in water waves. The basic equations here are the Euler equations and the equation of continuity for a non-viscous incompressible fluid moving under gravity. Throughout the book, in most considerations the motion is assumed to be irrotational and hence the existence of a velocity potential function is ensured in simply connected regions. In this case the equation of continuity for the velocity of the fluid is then reduced to the familiar Laplace equation for the velocity potential function.

Water waves are created normally by the presence of a free surface along which the pressure is constant. For the irrotational motion, on the free surface one than obtains the non-linear Bernoulli equation for the velocity potential function from the Euler equation. Based on small amplitude waves, linearised problems for the velocity potential function and for the free surface elevation are formulated.

At first we follow the derivation as can be found in [17, 19] to obtain equations for the wave potential in still water and as a superposition on a constant parallel flow potential. The coefficients in the free surface equations are constant. Then we derive linear equation for the superposition of small amplitude waves on a flow disturbed by some three dimensional object. If we consider the magnitude of the steady velocity vector to be small, we obtain for the time-dependent wave potential function a linear equation with non-constant coefficients.

1.1 Basic Linear Equations

The theory of water waves, to be presented here, is based on a model of non-viscous incompressible fluid moving under gravity. The equations of motion will be expressed in a right-handed system of rectangular coordinates x, y, z. In the Euler representation they read

$$u_t + uu_x + vu_y + wu_z = -\frac{1}{\rho} p_x,$$

A.J. Hermans, *Water Waves and Ship Hydrodynamics*,
DOI 10.1007/978-94-007-0096-3_1, © Springer Science+Business Media B.V. 2011

$$v_t + uv_x + vv_y + wv_z = -\frac{1}{\rho} p_y - g, \tag{1.1}$$

$$w_t + uw_x + vw_y + ww_z = -\frac{1}{\rho} p_z.$$

Here $u = u(x, y, z, t)$, $v = v(x, y, z, t)$, $w = w(x, y, z, t)$ are velocity components in the corresponding x, y, z direction; $p = p(x, y, z, t)$ is the pressure; ρ is the density of the fluid, a constant, and g is the gravitational acceleration. The continuity equation is

$$u_x + v_y + w_z = 0. \tag{1.2}$$

In most of the considerations the fluid motion is considered to be *irrotational*. This gives the additional set of equations

$$\begin{aligned} u_y - v_x &= 0, \\ v_z - w_y &= 0, \\ w_x - u_z &= 0, \end{aligned} \tag{1.3}$$

which guarantees in a simply connected region the existence of a velocity potential φ with

$$\begin{aligned} u &= \varphi_x, \\ v &= \varphi_y, \\ w &= \varphi_z. \end{aligned} \tag{1.4}$$

From (1.2) we see that φ satisfies Laplace's equation,

$$\varphi_{xx} + \varphi_{yy} + \varphi_{zz} = 0. \tag{1.5}$$

This greatly facilitates the theory.

In general, however, solutions of Laplace's equation will not show wave character, since the equation is elliptic. Waves are created by the presence of a free surface and are intimately related to the free surface condition.

1.2 Boundary Conditions

At the moving boundary the condition for a non-viscous fluid is very simple. The fluid velocity normal to the surface has to be equal to the normal component of the velocity of the surface itself. If the equation of the surface is given by

$$y = F(x, z, t), \tag{1.6}$$

we denote the velocity of a point on the surface by (U, V, W). A normal to the surface has the direction cosines

$$\left(\frac{F_x}{\sqrt{F_x^2+F_z^2+1}}, \frac{-1}{\sqrt{F_x^2+F_z^2+1}}, \frac{F_z}{\sqrt{F_x^2+F_z^2+1}}\right) \tag{1.7}$$

and the surface (or boundary) condition reads

$$\frac{uF_x - v + wF_z}{\sqrt{F_x^2+F_z^2+1}} = \frac{UF_x - V + WF_z}{\sqrt{F_x^2+F_z^2+1}} = \frac{-F_t}{\sqrt{F_x^2+F_z^2+1}}, \tag{1.8}$$

because

$$F_xU - V + F_zW + F_t = 0$$

for a point on the moving surface. Hence from (1.8) we have

$$v = F_t + uF_x + wF_z = \frac{\mathrm{d}F(x,z,t)}{\mathrm{d}t}, \tag{1.9}$$

which expresses the fact that, once a fluid particle is on the surface, it remains on the surface.

We will usually denote the bottom surface by $y = H(x,z,t)$, so that (1.9) reads

$$v = H_t + uH_x + wH_z. \tag{1.10}$$

Mostly in our considerations the bottom is fixed, that is H is independent of t, so that the term H_t in (1.10) vanishes.

The waves are created at the free surface, which is characterised by the condition that along this surface the pressure is a constant. Hence in addition to the kinematic equation

$$v = \eta_t + u\eta_x + w\eta_z, \tag{1.11}$$

for the free surface $y = \eta(x,z,t)$, we have the condition

$$p = \text{constant}, \tag{1.12}$$

along $y = \eta(x,z,t)$. There are two ways of formulating these conditions:

a. From the equations of motion (1.2), we find by inspection, in the case of irrotational motion, the Bernoulli equation

$$\varphi_t + \frac{1}{2}(u^2+v^2+w^2) + gy + \frac{p}{\rho} = f(t) \tag{1.13}$$

in which, because of the constant pressure, one can normalise φ to result in the dynamical free surface condition

$$\varphi_t + \frac{1}{2}(\varphi_x^2+\varphi_y^2+\varphi_z^2) + g\eta = \text{constant}. \tag{1.14}$$

b. The second way expresses that

$$\frac{\partial p}{\partial s_x} = 0, \qquad \frac{\partial p}{\partial s_z} = 0, \tag{1.15}$$

where s_x and s_z are coordinates on the free surface, which have their projections in the x and z directions, respectively. This gives[1]

$$\begin{aligned} \frac{\partial p}{\partial s_x} &= \frac{\partial p}{\partial x}\cos(x, s_x) + \frac{\partial p}{\partial y}\cos(y, s_x) = 0, \\ \frac{\partial p}{\partial s_z} &= \frac{\partial p}{\partial z}\cos(z, s_z) + \frac{\partial p}{\partial y}\cos(y, s_z) = 0 \end{aligned} \tag{1.16}$$

or

$$\begin{aligned} p_x + p_y\eta_x &= 0, \\ p_z + p_y\eta_z &= 0. \end{aligned} \tag{1.17}$$

Substituting (1.17) into (1.2), we have the relation

$$\begin{aligned} u_t + uu_x + vu_y + wu_z + \eta_x(v_t + uv_x + vv_y + wv_z) &= 0, \\ w_t + uw_x + vw_y + ww_z + \eta_z(v_t + uv_x + vv_y + wv_z) &= 0, \end{aligned} \tag{1.18}$$

which are also valid for rotational flow.

In this way the basic equations are derived. The further development of the theory is based on small parameter expansions of these equations. To do so an appropriate small dimensionless parameter has to be specified. Depending on the case considered, different formulations arise. In the next section we consider the case of a fixed bottom and where the water region is horizontally extended to infinity while no floating objects are present. This simplifies the theory considerably. Later we take other effects into account as well.

1.3 Linearised Theory

In this section we discuss two different cases, where we may obtain linearised equations for different situations. In the first one we assume that the waves are superimposed on a steady constant parallel flow field (current), while the second one deals with a wave field superimposed on a steady flow field, which obeys a simplified free surface condition. This steady flow may be generated by a slowly moving vessel. For fast moving objects one may need a more general non-linear theory for steady and unsteady boundary conditions. We will deal with some of these problems in future chapters.

---

[1] Note that $\cos(x, s_z) = \cos(z, s_x) = 0$.

1.3.1 Small Amplitude Waves in a Steady Current

The simplest approximation is the case where the deviation η of the free surface above a certain standard level, which is taken as $y = 0$, is small. We assume that

$$\eta(x, z, t) = \varepsilon\bar{\eta}(x, z, t), \tag{1.19}$$

where ε is a small dimensionless parameter. In addition we assume the bottom slope to be small of the same order of magnitude in ε and put

$$y = -h + \varepsilon h_1, \tag{1.20}$$

which will lead to the boundary condition

$$v = \varepsilon(h_{1t} + uh_{1x} + wh_{1z}) \tag{1.21}$$

from (1.10). For the free surface we obtain from (1.11) and (1.14)

$$v = \varepsilon(\bar{\eta}_t + u\bar{\eta}_x + w\bar{\eta}_z) \tag{1.22}$$

together with

$$\varphi_t + \frac{1}{2}(\varphi_x^2 + \varphi_y^2 + \varphi_z^2) + \varepsilon g\bar{\eta} = \text{constant}. \tag{1.23}$$

Now, for the solution of (1.5), we assume an expansion

$$\varphi(x, y, z, t) = \varphi_0 + \varepsilon\varphi_1 + \varepsilon^2\varphi_2 + \cdots, \tag{1.24}$$

and substitute it in (1.5) and boundary condition (1.21). Equating to zero the coefficients of like powers of ε, we get first that all φ_k's are harmonic functions. Moreover, we have from (1.20) and (1.21)

$$\left.\begin{aligned} v_0 &= \varphi_{0y} = 0, \\ v_1 &= h_{1t} + u_0 h_{1x} + w_0 h_{1z}, \end{aligned}\right\} \quad \text{at } y = -h. \tag{1.25}$$

Similarly, we expand $\bar{\eta}$ in (1.19) in the form

$$\bar{\eta} = \eta_1 + \varepsilon\eta_2 + \varepsilon^2\eta_2 + \cdots, \tag{1.26}$$

and find from (1.22) the free surface condition at $y = \varepsilon\bar{\eta}$,

$$\begin{aligned} &v_0 = 0 \quad \text{and} \\ &v_1 = \eta_{1t} + u_0\eta_{1x} + w_0\eta_{1z} \end{aligned} \tag{1.27}$$

together with

$$\begin{aligned} &\varphi_{0t} + \frac{1}{2}(\varphi_{0x}^2 + \varphi_{0y}^2 + \varphi_{0z}^2) = \text{constant}, \\ &\varphi_{1t} + u_0u_1 + v_0v_1 + w_0w_1 + g\eta_1 = 0 \end{aligned} \tag{1.28}$$

from (1.23).

The first approximation $\varphi_0, u_0 = \varphi_{0x}, v_0 = \varphi_{0y}, w_0 = \varphi_{0z}$, corresponds to a permanent flow. If we take the special case

$$u_0 = \text{constant},$$
$$v_0 = 0,$$
$$w_0 = \text{constant},$$

we can transform to a coordinate system with the x-axis in the direction of this constant flow and denote the velocity by U. In this case we have $\varphi_0 = Ux$ and the constant in (1.23) is equal to $\frac{1}{2}U^2$. Then we have the boundary condition from (1.25),

$$v_1 = Uh_{1x} \quad \text{at } y = -h, \tag{1.29}$$

and at the free surface $y = \varepsilon\bar{\eta}$, the coefficient of ε for (1.27) and (1.28) gives

$$\begin{aligned} \varphi_{1y} &= \eta_{1t} + U\eta_{1x}, \\ \varphi_{1t} &+ U\varphi_{1x} + g\eta_1 = 0. \end{aligned} \tag{1.30}$$

Instead of putting this condition (1.30) at $y = \varepsilon\bar{\eta}$, we put it at $y = 0$. Assuming that φ_1 admits an expansion in powers of $\varepsilon\bar{\eta}$, we then have

$$\begin{aligned} \varphi_{1x}(x, \varepsilon\bar{\eta}, z) &= \varphi_{1x}(x, 0, z) + \varepsilon\bar{\eta}\varphi_{1xy}(x, 0, z) + \cdots \\ &= \varphi_{1x}(x, 0, z) + \varepsilon\eta_1\varphi_{1xy}(x, 0, z) + O(\varepsilon^2), \end{aligned}$$

which leads to a modification of the terms of second order or higher. Hence the first approximation gives the following set of linear equations for φ_1 and η_1:

$$\begin{aligned} &\varphi_{1xx} + \varphi_{1yy} + \varphi_{1zz} = 0, \\ &\varphi_{1y} = h_{1t} + Uh_{1x} \qquad \text{at } y = -h, \\ &\left.\begin{aligned} &\varphi_{1y} = \eta_{1t} + U\eta_{1x} \\ &\varphi_{1t} + U\varphi_{1x} + g\eta_1 = 0 \end{aligned}\right\} \quad \text{at } y = 0. \end{aligned} \tag{1.31}$$

For a fixed flat bottom, h_1 is constant so that $h_{1x} = h_{1t} = 0$. For smooth functions, one can easily eliminate η_1 in the surface condition and obtain the formulation for the first-order approximation (dropping subscript 1):

$$\begin{aligned} &\varphi_{xx} + \varphi_{yy} + \varphi_{zz} = 0, \\ &\varphi_y = 0 \qquad \text{at } y = -h, \\ &U^2\varphi_{xx} + 2U\varphi_{xt} + \varphi_{tt} + g\varphi_y = 0 \quad \text{at } y = 0. \end{aligned} \tag{1.32}$$

Here the surface elevation η can be computed by

$$\eta = \frac{-1}{g}\left(\varphi_t + U\varphi_x\right). \tag{1.33}$$

Now given initial conditions, problems defined by (1.32) can be solved by means of the Laplace or Fourier transform method. As for illustration, we shall consider a few simple examples in Chap. 2.

1.3.2 Small Amplitude Waves in a Small Velocity Flow Field

Here we derive a free surface for unsteady waves superimposed on the steady free surface generated by a steady velocity field. This steady field may be generated by an object positioned in a constant parallel flow field. In general this leads to a very complicated condition, however if the magnitude of the velocity is small it can be simplified significantly. If no waves are present the magnitude of the velocity is characterised by a small non-dimensional Froude number $F = \frac{U}{\sqrt{gL}}$, where L is some length scale that plays a role in the problem, for instance the length of the disturbing object. It is assumed that this Froude number is small. In Sect. 5.2 we consider the diffraction of short waves if the steady flow field is generated by a parallel flow and is disturbed by a blunt object such as a sphere or a circular cylinder. In these cases we take for L the radius of the sphere or cylinder. Here we derive the free surface condition for such a case.

The easiest way is to follow the derivation, presented in Sect. 1.3.1, to determine a useful formulation for the steady potential. In this case of constant water depth the only small parameter is the Froude number $F = \frac{U}{\sqrt{gL}}$. Again we assume that the deviation of the free surface around $y = 0$ will be small. However we can not say that the free surface elevation is of $O(F)$. The order of magnitude of the elevation follows from the derivation and will turn out to be $O(F^2)$. For the steady case the kinematic free surface condition (1.11) becomes

$$v = u\eta_x + w\eta_z. \tag{1.34}$$

We assume that u, v and w are of the same order of magnitude $O(F)$. Hence for small values of η the kinematic condition reduces to

$$v = 0 \quad \text{at } y = 0. \tag{1.35}$$

The dynamic free surface condition now determines the order of magnitude of the corresponding free surface elevation. If we assume that in the far field the potential equals the unperturbed parallel flow Ux we obtain

$$\eta = \frac{-1}{2g}(u^2 + v^2 + w^2 - U^2). \tag{1.36}$$

Because of the specific form of the free surface condition (1.35) the steady potential described here is called the *double body* potential. For this potential we use the notation φ_r, the velocity components are written as $(u_r, v_r, w_r) = \nabla\varphi_r$ and the free surface elevation as η_r. If one is interested in the total steady potential one must derive an appropriate free surface condition also describing the *wavy pattern*. This will be done in Sect. 2.4. Our goal here is to derive a linearised free surface condition for the unsteady wave potential.

We assume that the potential φ can be decomposed as follows:

$$\varphi(x, y, z, t) = \varphi_r(x, y, z) + \varphi_0(x, y, z) + \varphi_w(x, y, z, t). \tag{1.37}$$

The potential φ_0 describes the steady wave pattern if waves are not present. Later we will show that this potential $\varphi_0 = o(\varphi_r)$, while as we have seen $\varphi_r = O(F)$. For this reason we neglect this term in the low Froude number small wave expansion and write

$$\varphi(x, y, z, t) = \varphi_r(x, y, z) + \varphi_w(x, y, z, t). \tag{1.38}$$

The free surface elevation $\eta(x, z, t)$ is assumed to be of the form

$$\eta(x, z, t) = \eta_r(x, z) + \eta_0(x, z) + \eta_w(x, z, t). \tag{1.39}$$

The function $\eta_0 = o(\eta_r)$, while $\eta_r = O(F^2)$, so we neglect η_0 and write

$$\eta(x, z, t) = \eta_r(x, z) + \eta_w(x, z, t). \tag{1.40}$$

We assume that the elevation of the free surface above the level $y = \eta_r(x, z)$ is small $O(\varepsilon)$. The condition for the wave potential at the bottom remains the same as before, however the free surface condition changes significantly. In principle the two small parameters are independent of each other. If the small Froude number is large compared with ε, we may introduce a new coordinate system $(x', y' - \eta_r(x', z'), z')$. The additional terms in the Laplace equation are small and may be neglected. The additional terms in the free surface condition may be neglected as well. If the two parameters are of the same order of magnitude we may linearise with respect to $y = 0$ directly, else it is defined at $y = \eta_r$. The kinematic condition as in (1.27), becomes

$$v_w = \eta_{wt} + u_r \eta_{wx} + w_r \eta_{wz}, \tag{1.41}$$

and if we use the surface condition the dynamic condition becomes

$$\varphi_{wt} + u_r \varphi_{wx} + w_r \varphi_{wz} + g\eta_w = 0. \tag{1.42}$$

We eliminate η_w by means of differentiation of (1.42) with respect to t, x and z respectively. The additional terms due to differentiation along the double body free surface η_r are $O(F^3)$ and may be neglected. For the wave potential we obtain the following formulation (we omit the primes):

$$\begin{aligned} &\varphi_{wxx} + \varphi_{wyy} + \varphi_{wzz} = 0, && \\ &\varphi_{wy} = 0 && \text{at } y = -h, \\ &\left(\frac{\partial}{\partial t} + u_r \frac{\partial}{\partial x} + w_r \frac{\partial}{\partial z}\right)^2 \varphi_w + g\frac{\partial \varphi_w}{\partial y} = 0 && \text{at } y = 0. \end{aligned} \tag{1.43}$$

The coefficients in the free surface condition depend on the local velocity. Although the formulation for the wave potential is linear, no simple solutions for a wave pattern can be given. In the case of the diffraction of short wave by a smooth object we will use an asymptotic wave theory. This method is developed in acoustic and electromagnetic theory, it is generally called the *ray method*. In Chap. 5 we present this asymptotic method.

Chapter 2
Linear Wave Phenomena

A few simple examples of the linearised boundary and initial-boundary value problems formulated in the previous chapter will be solved by the Fourier or Laplace transform method. Through these simple examples, basic wave phenomena or terminologies in water waves will be introduced. These are *phase velocity, dispersion relation, group velocity, wave fronts*, to name a few.

Of particular importance is the asymptotic behaviour of the free surface elevation for large values of relevant spaces and for time variables. This behaviour can be best obtained by the method of stationary phase (see Sect. 9.1). In this connection, the method of characteristics for treating first-order non-linear partial differential equations for the phase function is employed. Hence a brief summary of the concept of characteristics is included in Sect. 9.2.

A systematic derivation of oscillatory source singularity functions is presented for the disturbance below the free surface with and without current in Sects. 2.3 and 2.7.2. In Sect. 2.4 we derive for the steady case the field for a pressure disturbance at the free surface and for a point source below the free surface in Sect. 2.7.1. These source functions are often called Green functions and are used in numerical codes. One may derive different formulations for the functions as is shown.

2.1 Travelling Plane Waves

2.1.1 Plane Waves

It is easy to obtain travelling plane waves. As in Chap. 1 for small amplitude waves the linearised problem is defined by (1.32). For simplicity we restrict ourselves to the situation where $U = 0$. We consider two cases according to the water depth. We begin with the infinite depth. In this case the boundary value problem (1.32) consists of the Laplace equation

$$\varphi_{xx} + \varphi_{yy} + \varphi_{zz} = 0 \tag{2.1}$$

A.J. Hermans, *Water Waves and Ship Hydrodynamics*,
DOI 10.1007/978-94-007-0096-3_2, © Springer Science+Business Media B.V. 2011

together with the surface conditions

$$\varphi_{tt} + g\varphi_y = 0 \quad \text{at } y = 0 \tag{2.2}$$

and the condition at infinity

$$\varphi_y \to 0 \quad \text{as } y \to -\infty. \tag{2.3}$$

We seek a solution $\varphi(x, y, z, t)$ of (2.1)–(2.3) in the form

$$\varphi(x, y, z, t) = A\mathrm{e}^{\mathrm{i}(\alpha x+\beta z)+ky+\mathrm{i}\omega t}, \tag{2.4}$$

where α, β, k, ω and A are constants. Clearly (2.3) will be satisfied if k is positive. Substituting (2.4) into (2.1) and (2.2) we obtain

$$k = \alpha^2 + \beta^2 \quad \text{and} \quad -\omega^2 + gk = 0. \tag{2.5}$$

Set $\alpha = -k\cos\theta$ and $\beta = -k\sin\theta$ which clearly satisfy the first equation of (2.5) for any k. The second one gives that $k = \frac{\omega^2}{g}$ which is known as the *dispersion relation*—a relation between wave number k and frequency ω. Then the potential function has the form

$$\varphi(x, y, z, t) = A\exp\left\{-\mathrm{i}\omega\left[\frac{\omega}{g}(x\cos\theta + z\sin\theta) - t\right] + \frac{\omega^2}{g}y\right\}, \tag{2.6}$$

and consequently the water height is given by

$$\eta(x, z, t) = -\frac{1}{g}\varphi_t = -A\frac{\mathrm{i}\omega}{g}\exp\left\{-\mathrm{i}\omega\left[\frac{\omega}{g}(x\cos\theta + z\sin\theta) - t\right]\right\} \tag{2.7}$$

through use of (1.33). This formula represents plane waves.

For $\theta = 0$, we have plane waves travelling along the x-axis, independent of the z-coordinate:

$$\eta(x, t) = -\frac{\mathrm{i}\omega}{g}A\mathrm{e}^{-\mathrm{i}(\frac{\omega^2}{g}x-\omega t)} = A_1\mathrm{e}^{-\mathrm{i}\frac{\omega^2}{g}(x-ct)}, \tag{2.8}$$

where $c = \frac{g}{\omega}$ is the velocity of the wave (or *phase velocity*) and $A_1 = -\frac{\mathrm{i}\omega}{g}A$ is the amplitude of the wave. The real part of (2.8) corresponds to the real values wave height.

We now consider a wave train consisting of two plane waves in the x-direction with slightly different frequencies ω and $\omega + \delta\omega$. The total wave height may be written as

$$\begin{aligned}\eta(x, t) &= A_1\cos(kx - \omega t) + A_2\cos((k + \delta k)x - (\omega + \delta\omega)t \\ &= A(x, t)\cos(kx - \omega t + \theta(x, t))),\end{aligned} \tag{2.9}$$

where the amplitude function $A(x,t)$ and the phase function $\theta(x,t)$ are slowly varying functions. They can be written as

$$A(x,t)=\sqrt{A_1^2+A_2^2+2A_1A_2\cos(\delta kx-\delta\omega t)} \quad \text{and}$$
$$\tan\theta(x,t)=\frac{A_2\sin(\delta kx-\delta\omega t)}{A_1+A_2\cos(\delta kx-\delta\omega t)}. \tag{2.10}$$

The amplitude moves with the velocity $\frac{\delta\omega}{\delta k}$. It will be shown in Sect. 2.1.2 that the wave energy is proportional to the square of the amplitude, hence we may expect that the energy moves with a velocity

$$c_g=\lim_{\delta\omega\to 0}\frac{\delta\omega}{\delta k}=\frac{\mathrm{d}\omega}{\mathrm{d}k}. \tag{2.11}$$

This velocity c_g is called the *group velocity*.

The corresponding problem for finite water depth can be treated in the same way. We write

$$\varphi(x,y,z,t)=\hat{\varphi}(x,y,z)\mathrm{e}^{\mathrm{i}\omega t}.$$

Then in this case we have from (2.2) the surface condition

$$\hat{\varphi}_y=\frac{\omega^2}{g}\hat{\varphi} \quad \text{at } y=0, \tag{2.12}$$

while the condition at infinity (2.3) is replaced by the boundary condition (1.32). In terms of $\hat{\varphi}$ we have

$$\hat{\varphi}_y=0 \quad \text{at } y=-h. \tag{2.13}$$

For travelling waves in the direction of the x-axis, i.e., $\hat{\varphi}=\hat{\varphi}(x,y)$, a simple manipulation by the method of separation of variables leads to the solution

$$\varphi(x,y,t)=A\cosh[k(y+h)]\mathrm{e}^{-\mathrm{i}(kx-\omega t)}, \tag{2.14}$$

where the wave number k and the frequency ω are related by the dispersion relation

$$\omega^2=gk\tanh(kh). \tag{2.15}$$

Waves with a different wave number travel with a different phase velocity c which is defined by

$$c=\frac{\omega}{k}=\sqrt{\frac{g\tanh(kh)}{k}}. \tag{2.16}$$

Note that for kh small, since $\tanh(kh)=kh+\mathrm{O}((kh)^3)$, we have $c=\sqrt{gh}$ which is the case without dispersion. Observe again that if we let $h\to\infty$, we recover the case of infinite depth, (2.5). The dispersion causes a wave pattern, which at a certain place x and time t is a superposition of harmonic waves to be distorted at other places, because the components travel with different velocities. In the case of dispersion, it is difficult to determine the concept of 'wave speed'.

2.1.2 Wave Energy Transport

For the description of plane waves it is sufficient to restrict the considerations to the one-dimensional case. We represent at $t = 0$ the water height $\eta(x, 0)$ by the real integral

$$\eta(x) = \int_0^\infty C(k)\cos(kx)\,\mathrm{d}k + \int_0^\infty S(k)\sin(kx)\,\mathrm{d}k \tag{2.17}$$

with

$$C(k) = \frac{1}{\pi}\int_0^\infty \eta(x)\cos(kx)\,\mathrm{d}x, \quad \text{and}$$

$$S(k) = \frac{1}{\pi}\int_0^\infty \eta(x)\sin(kx)\,\mathrm{d}x.$$

Since $C(k)$ and $S(k)$ are respectively even and odd functions, setting

$$A(k) = \frac{1}{2}(C(k) + \mathrm{i}S(k)), \tag{2.18}$$

we can rewrite $\eta(x)$ as a complex integral

$$\eta(x) = \int_{-\infty}^\infty A(k)\mathrm{e}^{-\mathrm{i}kx}\,\mathrm{d}k. \tag{2.19}$$

A simple calculation shows that

$$\eta(x) = 2\Re\int_0^\infty A(k)\mathrm{e}^{-\mathrm{i}kx}\,\mathrm{d}k = \int_{-\infty}^\infty A^*(k)\mathrm{e}^{\mathrm{i}kx}\,\mathrm{d}k, \tag{2.20}$$

where $A^*(k)$ is the complex conjugate of $A(k)$.

For an understanding of the wave dispersion phenomenon, it is necessary to consider the energy propagation in the wave (linearised approximation). If the function $\eta(x)$ belongs to L^2, i.e., $\int_{-\infty}^\infty \eta(x)^2\,\mathrm{d}x$ exists, the potential energy is given by

$$E = \frac{1}{2}\rho g\int_{-\infty}^\infty \eta(x)^2\,\mathrm{d}x = \frac{1}{2}\int_{-\infty}^\infty\left(\int_{-\infty}^\infty A(k)\mathrm{e}^{-\mathrm{i}kx}\,\mathrm{d}k\right)\left(\int_{-\infty}^\infty A^*(k')\mathrm{e}^{\mathrm{i}k'x}\,\mathrm{d}k'\right)\mathrm{d}x$$

from (2.19) and (2.20). The latter integral can now be calculated by making use of the Fourier inversion theorem and the fact that $\int_{-\infty}^\infty \mathrm{e}^{\mathrm{i}(k'-k)x}\,\mathrm{d}x = 2\pi\delta(k'-k)$.

This gives

$$E = \frac{1}{2}\rho g 2\pi\int_{-\infty}^\infty |A(k)|^2\,\mathrm{d}k. \tag{2.21}$$

Hence from (2.18) we have

$$E = \frac{\rho g\pi}{4}\int_{-\infty}^\infty \{C(k)^2 + S(k)^2\}\,\mathrm{d}k. \tag{2.22}$$

If the dispersion relation $\omega = \omega(k)$ is known (for convenience we extend the definition of $\omega(-k) = -\omega(k)$), then we can compute the water height η at any arbitrary time t as follows:

$$\eta(x,t) = \int_{-\infty}^{\infty} A(k)\mathrm{e}^{\mathrm{i}(\omega t - kx)}\,\mathrm{d}k = \int_{-\infty}^{\infty} A(k)\mathrm{e}^{-\mathrm{i}(\omega t - kx)}\,\mathrm{d}k \tag{2.23}$$

in terms of the phase velocity $c = \omega/k$. Here it is assumed that the initial conditions are such that the wave propagates only in the projection of the positive x-axis.

The total potential energy is conserved; the wave only changes the distribution of the energy along the x-axis. In fact we have

$$\begin{aligned} E(t) &= \frac{1}{2}\rho g \int_{-\infty}^{\infty} |\eta(x,t)|^2\,\mathrm{d}x \\ &= \frac{1}{2}\rho g \int_{-\infty}^{\infty} \mathrm{d}x \left(\int_{-\infty}^{\infty} A(k)\mathrm{e}^{\mathrm{i}(\omega t - kx)}\,\mathrm{d}k\right)\left(\int_{-\infty}^{\infty} A^*(k')\mathrm{e}^{-\mathrm{i}(\omega' t - k'x)}\,\mathrm{d}k'\right) \end{aligned}$$

with $\omega' = \omega(k')$. The latter integral follows from (2.23) and can be calculated similarly according to the Fourier inversion theorem. We find again

$$E(t) = \rho g \pi \int_{-\infty}^{\infty} |A(k)|^2\,\mathrm{d}k. \tag{2.24}$$

Now we are going to find a measure for the velocity of the energy propagation and calculate to this end the location of the *centre of gravity* $\bar{x}(t)$ of the first moment of the energy, which is defined by

$$\bar{x}(t) = \frac{\int_{-\infty}^{\infty} x|\eta(x,t)|^2\,\mathrm{d}x}{\int_{-\infty}^{\infty} |\eta(x,t)|^2\,\mathrm{d}x}, \tag{2.25}$$

provided both integrals exist. Here the denominator has been shown to be a constant in time and can be calculated easily from (2.24). The numerator, however, requires some investigation. Equation (2.23) yields

$$\begin{aligned} &\int_{-\infty}^{\infty} x|\eta(x,t)^2|\,\mathrm{d}x \\ &\quad= \int_{-\infty}^{\infty} x\,\mathrm{d}x \int_{-\infty}^{\infty} A(k)\mathrm{e}^{\mathrm{i}(\omega t - kx)}\,\mathrm{d}k \int_{-\infty}^{\infty} A^*(k')\mathrm{e}^{-\mathrm{i}(\omega' t - k'x)}\,\mathrm{d}k' \\ &\quad= \mathrm{i}\int_{-\infty}^{\infty} \mathrm{d}x \int_{-\infty}^{\infty} A(k)\,\mathrm{d}\big(\mathrm{e}^{\mathrm{i}(\omega t - kx)}\big) \int_{-\infty}^{\infty} A^*(k')\mathrm{e}^{-\mathrm{i}(\omega' t - k'x)}\,\mathrm{d}k' \\ &\qquad+ t\int_{-\infty}^{\infty} \mathrm{d}x \int_{-\infty}^{\infty} A(k)\frac{\mathrm{d}\omega(k)}{\mathrm{d}k} \int_{-\infty}^{\infty} A^*(k')\mathrm{e}^{\mathrm{i}(\omega t - kx) - \mathrm{i}(\omega' t - k'x)}\,\mathrm{d}k\,\mathrm{d}k' \\ &\quad:= \mathscr{J}_1 + \mathscr{J}_2. \end{aligned}$$

Integrating by parts and taking account the fact that $A(k) \to 0$, as $k \to \pm\infty$ in view of Bessel's inequality, we find

$$\mathscr{J}_1 = -\mathrm{i}\int_{-\infty}^{\infty} \mathrm{d}x \int_{-\infty}^{\infty} \frac{\mathrm{d}A(k)}{\mathrm{d}k} \int_{-\infty}^{\infty} A^*(k')\mathrm{e}^{\mathrm{i}(\omega t - kx) - \mathrm{i}(\omega' t - k' x)}\,\mathrm{d}k\,\mathrm{d}k'.$$

Then by the Fourier inversion formula, we obtain

$$\mathscr{J}_1 = -2\pi\mathrm{i}\int_{-\infty}^{\infty} \frac{\mathrm{d}A(k)}{\mathrm{d}k} A^*(k)\,\mathrm{d}k;$$

$$\mathscr{J}_2 = 2\pi t\int_{-\infty}^{\infty} \frac{\mathrm{d}\omega(k)}{\mathrm{d}k} |A(k)|^2\,\mathrm{d}k.$$

Adding $\mathscr{J}_1$ and $\mathscr{J}_2$, we have

$$\int_{-\infty}^{\infty} x|\eta(x,t)^2|\,\mathrm{d}x = 2\pi\left\{-\mathrm{i}\int_{-\infty}^{\infty} \frac{\mathrm{d}A(k)}{\mathrm{d}k} A^*(k)\,\mathrm{d}k + t\int_{-\infty}^{\infty} \frac{\mathrm{d}\omega(k)}{\mathrm{d}k}|A(k)|^2\,\mathrm{d}k\right\}. \tag{2.26}$$

We define, as a mean value of a quantity $\psi(k)$ in the k-domain,

$$\bar{\psi} = \frac{\int_{-\infty}^{\infty} \psi(k)|A(k)|^2\,\mathrm{d}k}{\int_{-\infty}^{\infty} |A(k)|^2\,\mathrm{d}k}, \tag{2.27}$$

and remark that the first term in (2.26) determines the position of $\bar{x}(t)$ for $t = 0$. Hence we find

$$\bar{x}(t) = \bar{x}(0) + t\overline{\frac{\mathrm{d}\omega}{\mathrm{d}k}}, \tag{2.28}$$

i.e. the centre of gravity propagates with a velocity which is equal to the mean velocity of $\frac{\mathrm{d}\omega}{\mathrm{d}k}$. Here $\frac{\mathrm{d}\omega}{\mathrm{d}k}$ is called the *group velocity*; hence the mean value of the group velocity $\overline{\frac{\mathrm{d}\omega}{\mathrm{d}k}}$ is a measure for the speed of propagation of the energy.

The significance of this result becomes clear when we consider an amplitude spectrum $A(k)$, which extends only over a narrow wave number band:

$$\eta(x,t) = \int_{k_0-\varepsilon}^{k_0+\varepsilon} A(k)\mathrm{e}^{-\mathrm{i}(kx - \omega(k)t}\,\mathrm{d}k \quad \varepsilon > 0. \tag{2.29}$$

The centre of gravity satisfies

$$\bar{x}(t;k_0,\varepsilon) = \bar{x}(0;k_0,\varepsilon) + t\overline{\omega'(k_0,\varepsilon)}, \tag{2.30}$$

where $\overline{\omega'(k_0,\varepsilon)}$ is now the mean value of $\omega'(k) = \frac{\mathrm{d}\omega}{\mathrm{d}k}$ over the narrow band $[k_0 - \varepsilon, k_0 + \varepsilon]$. For small values of ε we simply replace $\overline{\omega'(k_0,\varepsilon)}$ by $\omega'(k_0)$.

For small values of t, one can make a more accurate analysis of the motion as follows. Expanding $\omega(k)$ in the form

$$\omega(k) = \omega(k_0) + (k - k_0)\omega'(k_0) + \frac{(k-k_0)^2}{2}\tilde{\omega}''(k),$$

and substituting into (2.29), we may write

$$\eta(x,t) = \int_{k_0-\varepsilon}^{k_0+\varepsilon} A(k)\mathrm{e}^{-\mathrm{i}\{(k_0x-\omega(k_0)t)+(k-k_0)(x-\omega'(k_0)t)-\frac{(k-k_0)^2}{2}\tilde{\omega}''(k)t\}}\,\mathrm{d}k$$

$$= \mathrm{e}^{-\mathrm{i}(k_0x-\omega(k_0)t)} \int_{k_0-\varepsilon}^{k_0+\varepsilon} A(k)\mathrm{e}^{-\mathrm{i}(k-k_0)(x-\omega'(k_0)t)}\,\mathrm{d}k + R, \qquad (2.31)$$

where

$$R = \mathrm{e}^{-\mathrm{i}(k_0x-\omega(k_0)t)} \int_{k_0-\varepsilon}^{k_0+\varepsilon} A(k)\mathrm{e}^{-\mathrm{i}(x-\omega'(k_0)t)(k-k_0)} \cdot \left\{\exp\left[\frac{\mathrm{i}(k-k_0)^2}{2}\tilde{\omega}''(k)t\right] - 1\right\}\mathrm{d}k.$$

Using the inequality $|\mathrm{e}^{\mathrm{i}u} - 1| \leq |u|$, we find an estimate of the remainder

$$|R| \leq \int_{k_0-\varepsilon}^{k_0+\varepsilon} |A(k)|\frac{(k-k_0)^2}{2}|\tilde{\omega}''(k)t|\,\mathrm{d}k$$

$$\leq \frac{1}{3}\Big(\max_{|k-k_0|<\varepsilon} |A(k)|\Big)\Big(\max_{|k-k_0|<\varepsilon} |\omega''(k)|\Big)\varepsilon^3 t,$$

which shows that for not too large values of t, the first term of (2.30) gives a good approximation of η. Assuming, for small ε, $A(k)$ to be constant $A(k_0)$ over the interval, we can integrate:

$$A(k_0)\mathrm{e}^{-\mathrm{i}(k_0x-\omega(k_0)t)} \int_{k_0-\varepsilon}^{k_0+\varepsilon} \mathrm{e}^{-\mathrm{i}(k-k_0)(x-\omega'(k_0)t)}\,\mathrm{d}k$$

$$= A(k_0)\mathrm{e}^{-\mathrm{i}k_0(x-\frac{\omega(k_0)}{k_0}t)}\frac{2\sin[(x-\omega'(k_0)t)\varepsilon]}{x-\omega'(k_0)t}.$$

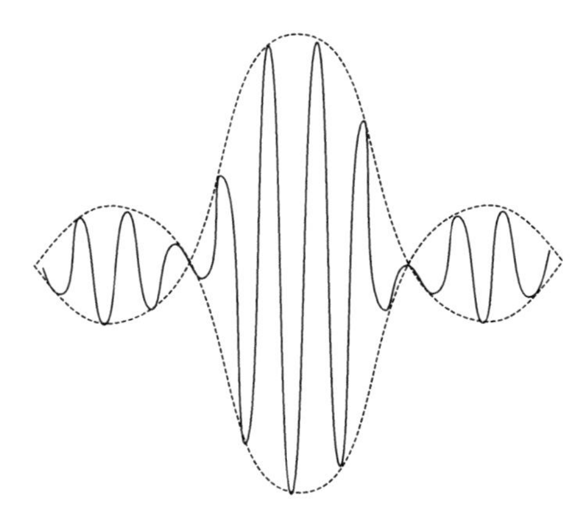

Fig. 2.1 Wave train

Hence we have, for small ε and t not too large,

$$\eta(x,t) \cong A(k_0)\mathrm{e}^{-\mathrm{i}k_0(x-\frac{\omega(k_0)}{k_0}t)}\frac{2\sin[(x-\omega'(k_0)t)\varepsilon]}{x-\omega'(k_0)t} \tag{2.32}$$

as shown in Fig. 2.1.

This represents a modulated wave; the amplitude moves with the group velocity $\omega'(k_)$ (the dotted enveloping curves) while the phase moves with the phase velocity $\omega(k_0)/k_0$ (the inscribed solid curves).

2.2 Cylindrical Waves

The boundary value problem for a cylindrical wave, at zero speed, $U=0$, is defined by the same equations in (1.32) for small amplitude waves. For harmonic oscillations we put

$$\begin{aligned} \varphi(x,y,z,t) &= \hat{\varphi}(x,y,z)\mathrm{e}^{\mathrm{i}\omega t}; \\ \eta(x,z,t) &= \hat{\eta}(x,z)\mathrm{e}^{\mathrm{i}\omega t}. \end{aligned} \tag{2.33}$$

Then the potential function $\hat{\varphi}(x,z,t)$ satisfies the Laplace equation

$$\hat{\varphi}_{xx}+\hat{\varphi}_{yy}+\hat{\varphi}_{zz}=0 \tag{2.34}$$

and the surface equation

$$\hat{\varphi}_y = \frac{\omega^2}{g}\hat{\varphi} \quad \text{for } y=0. \tag{2.35}$$

For infinite depth, we have again the condition

$$\hat{\phi} \text{ finite} \quad \text{for } y \to -\infty. \tag{2.36}$$

Since the problem now is axially symmetric, it is natural to make use of cylindrical coordinates $x=r\cos\theta$, $z=r\sin\theta$, $y=y$. Thus (2.34) reads

$$\frac{1}{r}\frac{\partial}{\partial r}\left(r\frac{\partial\hat{\varphi}}{\partial r}\right)+\frac{\partial^2\hat{\varphi}}{\partial y^2}=0. \tag{2.37}$$

We introduce dimensionless coordinates

$$\bar{r}=\frac{r\omega^2}{g}, \qquad \bar{y}=\frac{y\omega^2}{g}.$$

The transform leaves the differential equation (2.37) invariant, but the boundary condition (2.35) becomes

$$\hat{\varphi}=\hat{\varphi}_{\bar{y}} \quad \text{for } \bar{y}=0. \tag{2.38}$$

We solve this problem by the method of separation of variables and assume that

$$\hat{\varphi}(\bar{r}, \bar{y}) = \mathrm{e}^{\lambda \bar{y}} R(\bar{r}),$$

where $R(\bar{r})$ is a solution of the ordinary differential equation

$$\frac{1}{\bar{r}} \frac{\mathrm{d}}{\mathrm{d}\bar{r}} \left(\bar{r} \frac{\mathrm{d}R}{\mathrm{d}\bar{r}} \right) + \lambda^2 R = 0.$$

The boundary condition (2.38) gives that $\lambda = 1$. Thus we obtain

$$\hat{\varphi}(\bar{r}, \bar{y}) = \mathrm{e}^{\bar{y}} \left\{ A H_0^{(1)}(\bar{r}) + B H_0^{(2)}(\bar{r}) \right\}, \tag{2.39}$$

where $H_0^{(i)}$ are Hankel functions of order zero, and A and B are constants to be determined from the radiation condition as follows.

As is well known, for large values of $\bar{r}$ we have

$$H_0^{(1)}(\bar{r}) \approx \sqrt{\frac{2}{\pi \bar{r}}} \mathrm{e}^{\mathrm{i}(\bar{r} - \frac{\pi}{4})}, \quad \text{and}$$

$$H_0^{(2)}(\bar{r}) \approx \sqrt{\frac{2}{\pi \bar{r}}} \mathrm{e}^{-\mathrm{i}(\bar{r} - \frac{\pi}{4})}.$$

With time dependence $\mathrm{e}^{\mathrm{i}\omega t}$, only the solution

$$\varphi(\bar{r}, \bar{y}, \bar{t}) = B \mathrm{e}^{\bar{y}} H_0^{(2)}(\bar{r}) \mathrm{e}^{\mathrm{i}\omega t} \tag{2.40}$$

represents outgoing waves. For large values of $\bar{r}$ it behaves as

$$\varphi(\bar{r}, \bar{y}, \bar{t}) \approx B \mathrm{e}^{\bar{y}} \sqrt{\frac{2}{\pi \bar{r}}} \mathrm{e}^{-\mathrm{i}(\bar{r} - \omega t - \frac{\pi}{4})},$$

and the phase is defined by

$$\bar{r} - \omega t = \frac{\omega^2}{g} \left(r - \frac{g}{\omega} t \right),$$

which gives $\frac{g}{\omega}$ for the *phase velocity*.

The water height η is given by

$$\eta(\bar{r}, t) = -\frac{\mathrm{i}\omega}{g} B \mathrm{e}^{\bar{y}} H_0^{(2)}(\bar{r}) \mathrm{e}^{\mathrm{i}\omega t} \tag{2.41}$$

from (2.33), (1.33) and (2.40). Here it is understood that either the real or the imaginary part of the right-hand side of (2.41) is to be taken. We usually take the real part. This is an example of centred outgoing waves. The solution is obviously singular at $r = 0$ and $\forall y$. In the next Sect. 2.3 we will see that the far field of an harmonic point singularity has such a far-field behaviour.

2.3 Harmonic Source Singularity

It is of interest to determine the field disturbance of the free surface due to an harmonic singularity in a point below or at the free surface. As will be shown in Chap. 3, many methods to solve the problem of diffraction of waves by an object we make use of a distribution of singularities at the surface of the object. Here we will determine the field generated by such a singularity. As an example we treat the finite water depth case. The singularity is written as a *Dirac* δ-function in the right-hand side of the Laplace equation

$$\varphi_{xx} + \varphi_{yy} + \varphi_{zz} = \delta(x - x_0, y - y_0, z - z_0)\mathrm{e}^{\mathrm{i}\omega t}. \tag{2.42}$$

If we introduce $\varphi(x, y, z, t) = \hat{\varphi}(x, y, z)\mathrm{e}^{\mathrm{i}\omega t}$, the boundary value problem to be solved becomes

$$\begin{aligned} &\hat{\varphi}_{xx} + \hat{\varphi}_{yy} + \hat{\varphi}_{zz} = \delta(x - x_0, y - y_0, z - z_0), \\ &\hat{\varphi}_y = 0 \qquad \text{at } y = -h, \\ &\hat{\varphi}_y = \frac{\omega^2}{g}\hat{\varphi} \quad \text{at } y = 0. \end{aligned} \tag{2.43}$$

This formulation is not complete. We must add a condition at large horizontal distance from the source. The solution must fulfil the *radiation condition*. The disturbance for large values of $R = \sqrt{(x - x_0)^2 + (z - z_0)^2}$ may only consist of outgoing waves. The solution must have the form

$$\varphi(x, y, z, t) \approx A(R, y)\mathrm{e}^{-\mathrm{i}(kR - \omega t)}, \tag{2.44}$$

where the amplitude function tends to zero if $R \to 0$.

There are several ways to solve this problem. We shall employ the method of Fourier transform to obtain a solution. We introduce the following exponential transform of $\hat{\varphi}$ with respect to the x and z coordinates

$$\phi(\alpha, y, \beta) = \int_{-\infty}^{\infty}\int_{-\infty}^{\infty} \mathrm{e}^{\mathrm{i}(\alpha x + \beta z)}\hat{\varphi}(x, y, z)\,\mathrm{d}x\,\mathrm{d}z. \tag{2.45}$$

The inverse transform is

$$\hat{\varphi}(x, y, z) = \frac{1}{4\pi^2}\int_{-\infty}^{\infty}\int_{-\infty}^{\infty} \mathrm{e}^{-\mathrm{i}(\alpha x + \beta z)}\phi(\alpha, y, \beta)\,\mathrm{d}\alpha\,\mathrm{d}\beta. \tag{2.46}$$

We introduce the transform in the Laplace equation and the boundary conditions for $\hat{\varphi}$ and obtain an ordinary differential equation for ϕ with appropriate boundary conditions

$$\begin{aligned} &\phi_{yy} - (\alpha^2 + \beta^2)\phi = \mathrm{e}^{\mathrm{i}(\alpha x_0 + \beta z_0)}\delta(y - y_0), \\ &\phi_y = 0 \qquad \text{at } y = -h, \\ &\phi_y = \frac{\omega^2}{g}\phi \quad \text{at } y = 0. \end{aligned} \tag{2.47}$$

The singularity in the right-hand side of the differential equation can be replaced by the following conditions for the function $\phi(\alpha, \beta, y)$:

$$\lim_{\varepsilon \to 0} (\phi_y(\alpha, y_0 + \varepsilon, \beta) - \phi_y(\alpha, y_0 - \varepsilon, \beta)) = e^{i(\alpha x_0 + \beta z_0)},$$
$$\lim_{\varepsilon \to 0} (\phi(\alpha, y_0 + \varepsilon, \beta) - \phi(\alpha, y_0 - \varepsilon, \beta)) = 0. \tag{2.48}$$

The solution of the problem is written as ϕ^+ for $y_0 < y \leq 0$ and ϕ^- for $-h < y < y_0$. A convenient choice of the solution is

$$\phi^+ = A\cosh(k(y+h)) + B\sinh(k(y+h)),$$
$$\phi^- = C\cosh(k(y+h)).$$

Here k is defined as the distance to the origin in the Fourier space which is the positive root of $k^2 = \alpha^2 + \beta^2$. With this choice the bottom condition is fulfilled automatically. The constants A, B and C are determined by the condition at the free surface $y = 0$ together with the conditions at $y = y_0$. After some manipulations we find the solution for $y_0 < y \leq 0$,

$$\phi^+ = -\frac{\cosh(k(y_0+h))\{\nu \sinh(ky) + k\cosh(ky)\}}{k\{k\sinh(kh) - \nu\cosh(kh)\}} e^{i(\alpha x_0 + \beta z_0)}, \tag{2.49}$$

and for $-h < y < y_0$,

$$\phi^- = -\frac{\cosh(k(y+h))\{\nu \sinh(ky_0) + k\cosh(ky_0)\}}{k\{k\sinh(kh) - \nu\cosh(kh)\}} e^{i(\alpha x_0 + \beta z_0)}, \tag{2.50}$$

where $\nu = \frac{\omega^2}{g}$. We now apply the inverse transform given by (2.46) to ϕ^+

$$\hat{\varphi}^+(x, y, z) = \frac{-1}{4\pi^2} \int_{-\infty}^{\infty} \int_{-\infty}^{\infty} e^{-i(\alpha(x-x_0)+\beta(z-z_0))} \cdot \frac{\cosh(k(y_0+h))\{\nu \sinh(ky) + k\cosh(ky)\}}{k\{k\sinh(kh) - \nu\cosh(kh)\}} \, d\alpha \, d\beta. \tag{2.51}$$

It is convenient to introduce polar coordinates, both in the physical space and the Fourier space. We introduce

$$x - x_0 = R\cos\theta, \qquad z - z_0 = R\sin\theta \tag{2.52}$$

and

$$\alpha = k\cos\vartheta, \qquad \beta = k\sin\vartheta. \tag{2.53}$$

The solution can then be written as

$$\hat{\varphi}^+(x, y, z) = \frac{-1}{4\pi^2} \int_0^{2\pi} \int_0^{\infty} e^{-ikR\cos(\vartheta-\theta)} \cdot \frac{\cosh(k(y_0+h))\{\nu \sinh(ky) + k\cosh(ky)\}}{k\sinh(kh) - \nu\cosh(kh)} \, d\vartheta \, dk. \tag{2.54}$$

The integration with respect to ϑ can be carried out by making use of the following definition of the Bessel function J_0:

$$J_0(kR) = \frac{1}{2\pi}\int_0^{2\pi} \mathrm{e}^{-\mathrm{i}kR\cos(\vartheta-\theta)}\,\mathrm{d}\vartheta. \tag{2.55}$$

Hence, if we follow the same procedure for $\hat{\varphi}^-$, we obtain

$$\begin{aligned}\hat{\varphi}^+(x,y,z) &= \frac{-1}{2\pi}\int_0^{\infty} \frac{\cosh(k(y_0+h))\{\nu\sinh(ky)+k\cosh(ky)\}}{k\sinh(kh)-\nu\cosh(kh)} J_0(kR)\,\mathrm{d}k,\\ \hat{\varphi}^-(x,y,z) &= \frac{-1}{2\pi}\int_0^{\infty} \frac{\cosh(k(y+h))\{\nu\sinh(ky_0)+k\cosh(ky_0)\}}{k\sinh(kh)-\nu\cosh(kh)} J_0(kR)\,\mathrm{d}k.\end{aligned} \tag{2.56}$$

Until this point the radiation condition is not used. We will see that to define a proper inverse transform it has to be used. The integrands of the functions $\hat{\varphi}^{+,-}$ each have a singularity for a real value of the denominator. Hence, the integrals are not well defined. The equation $k\sinh(kh) - \nu\cosh(kh) = 0$ has one real root together with an infinite number of purely imaginary roots. From the theory of Fourier integral we know that the contour of integration has to pass, in the complex k-plane, above or below the singularity. The choice is determined by the radiation condition. A way to determine the correct choice is to introduce a small artificial damping in the fluid. If we assume the far field to be of the form $\mathrm{e}^{-\mathrm{i}(kR-\omega t)}$ we see that the only choice for vanishing waves is to introduce a complex wave number of the form $k = \bar{k} - \mathrm{i}\tilde{k}$. The negative imaginary part of the wave number may be generated by some artificial, non-physical, damping. This indicates that the singularity on the real axis must be passed above. Representation (2.56) for $\hat{\varphi}$ consists of different forms depending on whether y is larger or smaller than y_0. This might be not practical. One may obtain a single expression if we use some lemmas from the theory of complex functions. We use the following lemma for analytic functions $f(z)$ and $g(z)$, while the function $f(z)$ has simple zeros z_i in the complex plane. If we define $f(z) = z\sinh(zh) - \nu\cosh(zh)$ and $g(z) = \cosh(zp)\{\nu\sinh(zq) + z\cosh(zq)$ respectively, then for $|z| \to \infty$ the function $\frac{g(z)}{f(z)} \to 0$ fast enough and we have

$$\frac{g(z)}{f(z)} = \frac{g(0)}{f(0)} + \sum_i g(z_i)\gamma_i\left(\frac{1}{z-z_i} + \frac{1}{z_i}\right) \quad \text{with } \gamma_i = \frac{1}{f_z(z_i)}, \tag{2.57}$$

which is an expansion of $\frac{g(z)}{f(z)}$ in rational fractions of z, see [21], Sect. 7.4.

The integrands of both integrals in the expression for $\hat{\varphi}(x,y,z)$ (2.56) has infinitely many simple poles $k = \pm k_i$ $(i = 0, 1, 2, \ldots)$ in the complex k-plane. We have

$$k_i\sinh(k_i h) - \nu\cosh(k_i h) = 0. \tag{2.58}$$

The positive real zero is k_0, while the positive imaginary roots are $k_i = \mathrm{i}\kappa_i$ $(i = 1, 2, \ldots)$, see Fig. 2.2.

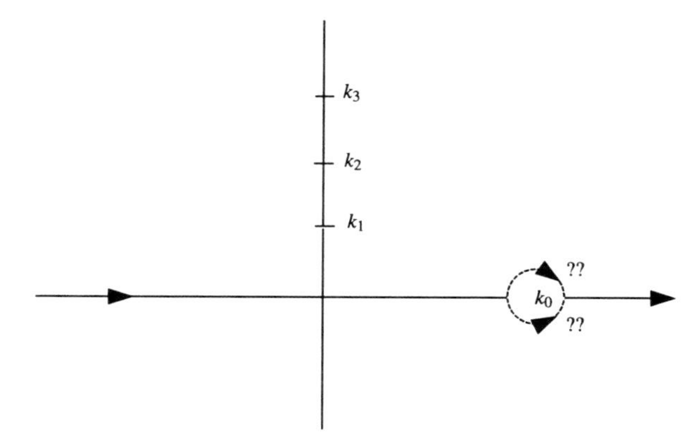

Fig. 2.2 The singularities in the complex k-plane

According to (2.57) we may write

$$\frac{g(k)}{k\sinh(kh) - \nu\cosh(kh)} = \sum_{i=0}^{\infty} g(k_i)\left(\frac{\alpha_i^+}{k-k_i} + \frac{\alpha_i^-}{k+k_i}\right), \tag{2.59}$$

where we used the fact that in our case $g(k)$ is antisymmetric and $g(0) = 0$ and where α_i is defined as

$$\alpha_i^{\pm} = \frac{\pm k_i}{(\nu + k_i^2 h - \nu^2 h)\cosh(k_i h)}. \tag{2.60}$$

If we work out the integrands of (2.56) we find one expression for $\hat{\varphi}(x, y, z)$, valid for $-h < y \leq 0$. We obtain

$$\hat{\varphi}(x, y, z) = \frac{-1}{2\pi}\sum_{i=0}^{\infty} \frac{k_i^2 - \nu^2}{\nu + k_i^2 h - \nu^2 h} \cosh(k_i(y+h))\cosh(k_i(y_0+h)) \cdot \int_0^{\infty}\left(\frac{1}{k-k_i} - \frac{1}{k+k_i}\right) J_0(kR)\,\mathrm{d}k. \tag{2.61}$$

The integral in the right hand side can, by introducing $k = -k^*$ in the second part, be rewritten as

$$\mathscr{I}(k_i) = \frac{1}{2}\int_{-\infty}^{\infty} \frac{H_0^{(1)}(kR)}{k-k_i}\,\mathrm{d}k + \frac{1}{2}\int_{-\infty}^{\infty} \frac{H_0^{(2)}(kR)}{k-k_i}\,\mathrm{d}k. \tag{2.62}$$

Due to the asymptotic behaviour of the Hankel functions we may close the first integral in the upper half of the complex k plane, while the second one may be closed in the lower half. In this way the contributions of the contours at $|k| \to \infty$ tend to zero. If the path of integration in (2.62) passes the singularity $k = k_0$ in the upper plane we obtain the following result for $i = 0$:

$$\mathscr{I}(k_0) = -\pi \mathrm{i} H_0^{(2)}(k_0 R), \tag{2.63}$$

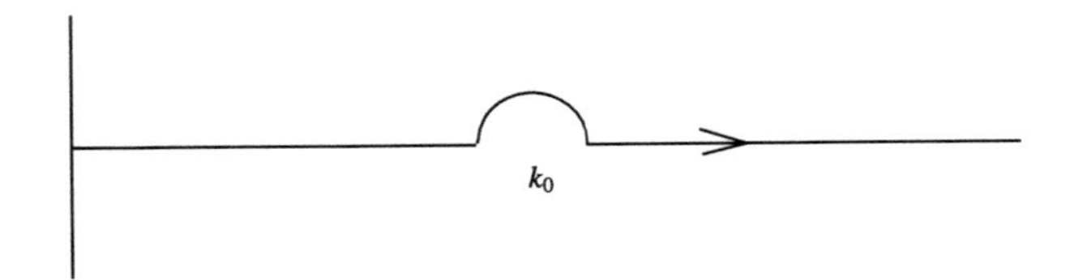

Fig. 2.3 Line of integration

and for $i = 1, 2, \ldots$

$$\mathscr{J}(k_i) = \pi \mathrm{i} H_0^{(1)}(\mathrm{i}\kappa_i R) = 2K_0(\kappa_i R), \tag{2.64}$$

where $K_0(z)$ is the modified Bessel function. The contribution of $H_0^{(2)}(k_0 R)$ represents an outgoing circular wave, while the contribution of each $K_0(\kappa_i R)$ is exponential decaying for large values of R. This confirms the right choice of the contour of integration, see Fig. 2.3. We notice that the use of an artificial damping to shift k_0 actually is not the only way to find the correct contour of integration. If one chooses the contour to pass underneath k_0 the wavy behaviour is described by $H_0^{(1)}(k_0 R)$, describing an incoming circular wave field. Waves travelling towards the source clearly which disobey the radiation condition.

The expression for the total field now becomes

$$\varphi(x, y, z, t) = \mathrm{e}^{\mathrm{i}\omega t} \hat{\varphi}(x, y, z)$$

with

$$\begin{aligned}\hat{\varphi}(x, y, z) = & \frac{\mathrm{i}(k_0^2 - \nu^2)}{2(\nu + k_0^2 h - \nu^2 h)} \cosh(k_0(y+h)) \cosh(k_0(y_0+h)) H_0^{(2)}(k_0 R) \\ & + \frac{1}{\pi} \sum_{i=1}^{\infty} \frac{\kappa_i^2 + \nu^2}{\nu - \kappa_i^2 h - \nu^2 h} \cos(\kappa_i(y+h)) \cos(\kappa_i(y_0+h)) \\ & \times K_0(\kappa_i R).\end{aligned} \tag{2.65}$$

If we take the real part of (2.65) and multiply it with -4π we have the famous result of F. John. The different factor originates from the normalisation of the point source. This formulation can be used to compute the disturbance due to a unit point source at finite difference from the source. However, the series does not converge close to the source. This was to be expected, because of the singular, $\frac{-1}{4\pi r}$, behaviour of $\hat{\varphi}$, where $r = \sqrt{(x-x_0)^2 + (y-y_0)^2 + z - z_0)^2}$ is the distance to the singularity.

We expect to find a useful solution near the singularity if we write it as

$$\hat{\varphi}(x, y, z) = -\frac{1}{4\pi r} - \frac{1}{4\pi \tilde{r}} + \psi(x, y, z), \tag{2.66}$$

where $\tilde{r} = \sqrt{(x-x_0)^2 + (y+2h+y_0)^2 + (z-z_o)^2}$ is the distance to the mirror image, with respect to the bottom, of the source point. For $\psi(x, y, z)$ we obtain the following problem:

$$
\begin{aligned}
&\psi_{xx} + \psi_{yy} + \psi_{zz} = 0, \\
&\psi_y = 0 && \text{at } y = -h, \\
&\psi_y - \nu\psi = \frac{1}{4\pi}\left\{\frac{\partial}{\partial y}\left(\frac{1}{r} + \frac{1}{\tilde{r}}\right) - \nu\left(\frac{1}{r} + \frac{1}{\tilde{r}}\right)\right\} && \text{at } y = 0. \\
&\qquad := g(x, z; x_0, y_0, z_0)
\end{aligned}
\tag{2.67}
$$

We apply the double Fourier transform to the function ψ,

$$
\Psi(\alpha, y, \beta) = \int_{-\infty}^{\infty}\int_{-\infty}^{\infty} \mathrm{e}^{\mathrm{i}(\alpha x + \beta z)}\psi(x, y, z)\,\mathrm{d}x\,\mathrm{d}z \tag{2.68}
$$

and introduce polar coordinates (2.56) in the Fourier space. The ordinary differential equation and boundary conditions for Ψ become

$$
\begin{aligned}
&\Psi_{yy} - k^2\Psi = 0, \\
&\Psi_y = 0 && \text{at } y = -h, \\
&\Psi_y - \nu\Psi = G(k; x_0, y_0, z_0) && \text{at } y = 0.
\end{aligned}
\tag{2.69}
$$

We make use of the known transform of $\frac{-1}{4\pi r}$, the point source for an infinite fluid where no free surface is present

$$
\mathscr{F}\left(\frac{-1}{4\pi r}\right) = \frac{-1}{2k}\mathrm{e}^{\mathrm{i}(\alpha x_0 + \beta z_0) - k|y - y_0|}. \tag{2.70}
$$

This formula can be obtained by means of the double Fourier transform to the Laplace equation, as before, in the case of an infinite fluid. If we apply this formula to $g(x, z; x_0, y_0, z_0)$ we obtain

$$
G(k; x_0, y_0, z_0) = -\frac{k + \nu}{k}\mathrm{e}^{-kh}\cosh(k(y_0 + h))\mathrm{e}^{\mathrm{i}(\alpha x_0 + \beta z_0)} \tag{2.71}
$$

and the solution of (2.69) becomes

$$
\Psi(\alpha, y, \beta) = -\frac{k + \nu}{k}\mathrm{e}^{-kh}\frac{\cosh(k(y + h))\cosh(k(y_0 + h))}{k\sinh(kh) - \nu\cosh(kh)}\mathrm{e}^{\mathrm{i}(\alpha x_0 + \beta z_0)}. \tag{2.72}
$$

The inverse Fourier transform is defined as

$$
\psi(x, y, z) = \frac{1}{4\pi^2}\int_{-\infty}^{\infty}\int_{-\infty}^{\infty}\Psi(\alpha, y, \beta)\mathrm{e}^{-\mathrm{i}(\alpha x + \beta z)}\,\mathrm{d}\alpha\,\mathrm{d}\beta. \tag{2.73}
$$

With the introduction of polar coordinates in the physical (2.52) and Fourier (2.53) space we obtain with the use of (2.55) the total field

$$
\varphi(x, y, z, t) = \mathrm{e}^{\mathrm{i}\omega t}\hat{\varphi}(x, y, z)
$$

with

$$\hat{\varphi}(x,y,z) = -\frac{1}{4\pi r} - \frac{1}{4\pi \tilde{r}} - \frac{1}{2\pi}\int_0^\infty \frac{(k+\nu)\mathrm{e}^{-kh}\cosh(k(y+h))\cosh(k(y_0+h))}{k\sinh(kh)-\nu\cosh(kh)} J_0(kR)\,\mathrm{d}k. \tag{2.74}$$

If we introduce some artificial damping in the problem we observe that the contour of integration passes above the real pole in the integrand. This finally leads to the expression

$$\begin{aligned}\hat{\varphi}(x,y,z) &= -\frac{1}{4\pi r} - \frac{1}{4\pi \tilde{r}} \\ &\quad - \frac{1}{2\pi} ⨍_0^\infty \frac{(k+\nu)\mathrm{e}^{-kh}\cosh(k(y+h))\cosh(k(y_0+h))}{k\sinh(kh)-\nu\cosh(kh)} J_0(kR)\,\mathrm{d}k \\ &\quad + \frac{\mathrm{i}}{2}\frac{(k_0+\nu)\mathrm{e}^{-k_0 h}\sinh(k_0 h)\cosh(k_0(y+h))\cosh(k_0(y_0+h))}{\nu h + \sinh^2(k_0 h)} J_0(k_0 R),\end{aligned} \tag{2.75}$$

where $⨍$ indicates the principal value of the integral. If we are interested in the deep water case we may obtain an expression for the source potential by using (2.75) for large values of h. We obtain for the limit $h \to \infty$,

$$\hat{\varphi}(x,y,z) = -\frac{1}{4\pi r} - \frac{1}{4\pi} ⨍_0^\infty \frac{k+\nu}{k-\nu}\mathrm{e}^{k(y+y_0)} J_0(kR)\,\mathrm{d}k + \frac{\mathrm{i}}{2}\nu \mathrm{e}^{\nu(y+y_0)} J_0(\nu R). \tag{2.76}$$

This result may be rewritten as

$$\hat{\varphi}(x,y,z) = -\frac{1}{4\pi r} + \frac{1}{4\pi \bar{r}} - \frac{1}{4\pi} ⨍_0^\infty \frac{2k}{k-\nu}\mathrm{e}^{k(y+y_0)} J_0(kR)\,\mathrm{d}k + \frac{\mathrm{i}}{2}\nu \mathrm{e}^{\nu(y+y_0)} J_0(\nu R), \tag{2.77}$$

where $\bar{r} = \sqrt{(x-x_0)^2 + (y+y_0)^2 + (z-z_0)^2}$ is the distance to the mirror point, with respect to the unperturbed free surface.

The contour of integration may be deformed to obtain different forms of (2.74). We can rewrite the integral as a contribution of the pole and an integral along the vertical axis of the complex k-plane. Instead of the way the solution is written in (2.76) one also may write the solution as the sum $-(\frac{1}{4\pi r} + \frac{1}{4\pi \bar{r}})$, where $-\frac{1}{4\pi \bar{r}}$ is the field of a singularity located at $(x_0, -y_0, z_0)$ in an infinite fluid, and use an integral expression for this term. There are more choices possible, they are sometimes used in the literature for different reasons.

2.4 The Moving Pressure Point

We consider the field generated by a pressure point disturbance at the free surface, moving in the direction of the positive x-axis. For small amplitude waves the linearised free surface condition is defined by (1.32). We suppose the bottom at infinity, $y = -\infty$. Hence the bottom condition is replaced by the condition that φ remains finite as $y \to -\infty$. We look for a very simple solution in a steady flow, for which everywhere at $y = 0$ except at $x = z = 0$ the pressure vanishes. By introducing the dimensionless coordinates

$$\bar{x} = \frac{xg}{U^2}, \qquad \bar{y} = \frac{yg}{U^2}, \qquad \bar{z} = \frac{zg}{U^2},$$

we can formulate the boundary value problem as follows;

$$\begin{aligned} &\varphi_{\bar{x}\bar{x}} + \varphi_{\bar{y}\bar{y}} + \varphi_{\bar{z}\bar{z}} = 0, \\ &\varphi_{\bar{x}\bar{x}} + \varphi_{\bar{y}} = 0 \quad \text{at } \bar{y} = 0, (\bar{x}, \bar{z}) \neq (0, 0), \\ &\varphi \text{ finite} \qquad \text{as } \bar{y} \to \infty. \end{aligned} \tag{2.78}$$

We seek solutions of (2.78) by means of a Fourier transform with respect to $\bar{x}$,

$$\hat{\varphi}(\alpha, \bar{y}, \bar{z}) = \int_{-\infty}^{\infty} \mathrm{e}^{\mathrm{i}\alpha\bar{x}} \varphi(\bar{x}, \bar{y}, \bar{z})\, \mathrm{d}\bar{x} \tag{2.79}$$

with its inverse transform

$$\varphi(\bar{x}, \bar{y}, \bar{z}) = \frac{1}{2\pi} \int_{-\infty}^{\infty} \mathrm{e}^{-\mathrm{i}\alpha\bar{x}} \hat{\varphi}(\alpha, \bar{y}, \bar{z})\, \mathrm{d}\bar{\alpha}. \tag{2.80}$$

This leads to the boundary value problem for $\hat{\varphi}(\alpha, \bar{y}, \bar{z})$:

$$\begin{aligned} &\hat{\varphi}_{\bar{y}\bar{y}} + \hat{\varphi}_{\bar{z}\bar{z}} - \alpha^2 \hat{\varphi} = 0, \\ &-\alpha^2 \hat{\varphi} + \hat{\varphi}_{\bar{y}} = 0 \quad \text{at } \bar{y} = 0, \\ &\hat{\varphi} \text{ finite} \qquad \text{as } \bar{y} \to -\infty. \end{aligned} \tag{2.81}$$

A simple solution of (2.81) can be found in the form

$$\hat{\varphi} = \mathrm{e}^{\alpha^2 \bar{y}} F(\bar{z}),$$

where $F(\bar{z})$ satisfies the equation

$$(\alpha^4 - \alpha^2) F + F_{\bar{z}\bar{z}} = 0.$$

Consequently, we take as a possible solution

$$\varphi(\bar{x}, \bar{y}, \bar{z}) = \frac{A}{2\pi} \int_{-\infty}^{\infty} \exp\{-\mathrm{i}\alpha\bar{x} + \alpha^2 \bar{y} + \mathrm{i}\alpha(\alpha^2 - 1)^{\frac{1}{2}} \bar{z}\}\, \mathrm{d}\alpha, \tag{2.82}$$

for A being a constant. Note that $\varphi(\bar{x}, \bar{y}, \bar{z})$ is not defined for $\bar{x} = \bar{y} = \bar{z} = 0$. From (1.33) we find the free surface elevation

$$\eta(\bar{x}, \bar{z}) = \frac{A\mathrm{i}}{2\pi U} \lim_{\bar{y}\to 0} \int_{-\infty}^{\infty} \left(\alpha \mathrm{e}^{\alpha^2 \bar{y}}\right) \exp\left\{\mathrm{i}\left(-\alpha\bar{x} + \alpha(\alpha^2 - 1)^{\frac{1}{2}}\bar{z}\right)\right\} \mathrm{d}\alpha \tag{2.83}$$

which apparently is infinite for $\bar{x} = \bar{z} = 0$.

In order to get a better insight into the shape of the surface we shall evaluate this expression (2.83) for large values of $\bar{x}$ and $\bar{z}$; that is distances to the origin that are large compared to the reference length U^2/g. This evaluation is performed by the method of stationary phase (see Sect. 9.1).

We note that if we let $R = (\bar{x}^2 + \bar{z}^2)^{\frac{1}{2}}$, $\bar{x} = R\cos\vartheta$ and $\bar{z} = R\sin\vartheta$, then for each fixed ϑ, (2.83) can be written in the form

$$\int_{-\infty}^{\infty} g(\alpha) \exp(\mathrm{i}Rf(\alpha))\, \mathrm{d}\alpha,$$

where

$$g(\alpha) := \frac{A\mathrm{i}}{2\pi U}\alpha \quad \text{and}$$

$$Rf(\alpha) := -\alpha\bar{x} + \alpha(\alpha^2 - 1)^{\frac{1}{2}}\bar{z}.$$

Hence the stationary points are solutions of the equation

$$\frac{\partial}{\partial\alpha}\left\{-\alpha\bar{x} + \alpha(\alpha^2 - 1)^{\frac{1}{2}}\bar{z}\right\} = 0. \tag{2.84}$$

(cf. (9.13)).

Let α_0 be a solution of (2.84). We obtain therefore the asymptotic form of $\eta(\bar{x}, \bar{z})$:

$$\eta(\bar{x}, \bar{z}) \cong \frac{A\mathrm{i}}{\pi U}\alpha_0 \sqrt{\frac{\pi\,\mathrm{i}\alpha_0(\alpha_0^2 - 1)^{\frac{3}{2}}}{2\bar{z}(2\alpha_0^2 - 3)}} \exp\left\{\mathrm{i}\left(-\alpha_0\bar{x} + \alpha_0(\alpha_0^2 - 1)^{\frac{1}{2}}\bar{z}\right)\right\} \tag{2.85}$$

The phase function is of the most importance. If we put

$$\psi = -\alpha_0\bar{x} + \alpha_0(\alpha_0^2 - 1)^{\frac{1}{2}}\bar{z}, \tag{2.86}$$

we obtain from (2.84)

$$-\bar{x} + \frac{2\alpha_0^2 - 1}{\sqrt{\alpha_0^2 - 1}}\bar{z} = 0. \tag{2.87}$$

Fig. 2.4 Kelvin pattern

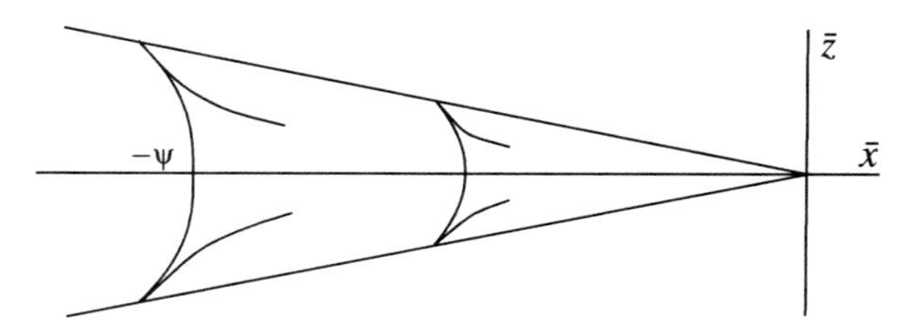

Setting $\alpha_0 = \frac{1}{\cos\theta}$, we obtain from (2.86) and (2.87) the equations

$$\begin{aligned} \bar{x} &= -\psi(2\cos\theta - \cos^3\theta) = -\frac{1}{4}\psi(5\cos\theta - \cos(3\theta)), \\ \bar{z} &= -\psi\cos^2\theta\sin\theta = -\frac{1}{4}\psi(\sin\theta + \sin(3\theta)) \end{aligned} \tag{2.88}$$

for the curves of constant phase ψ, which give the wave pattern. These curves are all similar with the origin as centre, and have wave cusps at $\bar{x} = \bar{z} = 0$ (or $\theta = \pi/2$) and at the points where $\frac{d\bar{x}}{d\theta} = \frac{d\bar{z}}{d\theta} = 0$. Since

$$\frac{d\bar{x}}{d\theta} = -\psi\sin\theta(2 - 3\cos^2\theta) \quad \text{and} \quad \frac{d\bar{z}}{d\theta} = -\psi\cos\theta(3\cos^2\theta - 2),$$

it follows that at the points

$$\bar{x} = -\psi\frac{4\sqrt{6}}{9}, \qquad \bar{z} = -\psi\frac{2\sqrt{3}}{9}$$

corresponding to $\cos\theta = \sqrt{2/3}$, there are cusps (it is understood that the expression (2.85) is not valid in the neighbourhood of cusps). A typical curve is shown in Fig. 2.4. We see that the curve intersects the $\bar{x}$-axis at the points $\bar{x} = -\psi$ (corresponding to $\theta = 0$). The cusps lie on a straight line, through the origin, which makes a fixed angle with the $\bar{x}$-axis. The pattern obtained this way is called the *Kelvin* wave pattern.

2.5 Wave Fronts

In view of (2.7) and (2.23), we now consider the general representation for the free surface elevation:

$$\eta(x, z, t) = \int_{-\infty}^{\infty} A(k)e^{-i(kx\cos\theta + kz\sin\theta - \omega t)}\,dk. \tag{2.89}$$

In particular, we are interested in the asymptotic behaviour of η for large values of t. We apply the method of stationary phase, see Sect. 9.1, to (2.89). The method requires the determination of the value of k for which the phase

$$\Psi(x, z, k) = -\omega t + k(x\cos\theta + z\sin\theta)$$

$$= -t\left[k\left(\frac{x}{t}\cos\theta + \frac{z}{t}\right) - \omega(k)\right] \tag{2.90}$$

is stationary. (Here we consider Ψ as depending on the three parameters $\frac{x}{t}$, $\frac{z}{t}$ and t. For each pair of values of $\frac{x}{t}$ and $\frac{z}{t}$, the asymptotic expansion for η is considered for large t.) This leads to the consideration of solutions of the equations

$$\frac{d\Psi}{dk} = 0 \quad \text{or} \quad -\frac{d\omega}{dk}t + x\cos\theta + z\sin\theta = 0. \tag{2.91}$$

Let k_0 be any solution of (2.91). Then the approximate result for large t is

$$\eta(x,z,t) = A(k_0)\sqrt{\frac{2\pi}{t|\omega''(k_0)|}}\mathrm{e}^{-\mathrm{i}(k_0 x\cos\theta + k_0 z\sin\theta - \omega(k_0)t - \frac{\pi}{4}\operatorname{sgn}\omega''(k_0))} \tag{2.92}$$

provided that $\frac{d^2\Psi(k_0)}{dk^2} \neq 0$, i.e. $\omega''(k_0) \neq 0$.

The lines $\Psi =$ constant are lines of constant phase; these lines are called *wave fronts*. We can define a partial differential equation for the wave fronts from the dispersion relation $\omega = H(k)$. In fact, we can express k_0 in terms of x, z, t and θ from (2.92) so that differentiations of (2.91) (with $k = k_0$) with respect to these variables yield

$$\begin{aligned}
\Psi_x &= k_0\cos\theta + (x\cos\theta + z\sin\theta - \omega_0' t)\frac{\partial k_0}{\partial x} = k_0\cos\theta, \\
\Psi_z &= k_0\sin\theta + (x\cos\theta + z\sin\theta - \omega_0' t)\frac{\partial k_0}{\partial z} = k_0\sin\theta, \\
\Psi_t &= -\omega_0 + (x\cos\theta + z\sin\theta - \omega_0' t)\frac{\partial k_0}{\partial t} = -\omega_0,
\end{aligned} \tag{2.93}$$

with $\omega_0 = H(k_0)$. The first two equations of (2.93) imply that

$$k_0^2 = \Psi_x^2 + \Psi_z^2$$

with which the third one shows that the dispersion relation $\omega_0 = H(k_0)$ gives a partial differential equation for the phase function Ψ, the Hamilton-Jacobi equation

$$\begin{aligned}
&\Psi_t + H\left(\sqrt{\Psi_x^2 + \Psi_y^2}\right) = 0 \quad \text{or} \\
&\Psi_t + H\left(\sqrt{p^2 + q^2}\right) = 0,
\end{aligned} \tag{2.94}$$

where $p = \Psi_x = k_0\cos\theta$ and $q = \Psi_z = k_0\sin\theta$ are the conjugate variables to x and z, respectively. We have just seen that the wave fronts correspond to level curves of the Hamilton-Jacobi equation. But in wave phenomena one expects the dual concept of rays to appear also. The rays in the present case are the characteristics of the

Fig. 2.5 Wave fronts

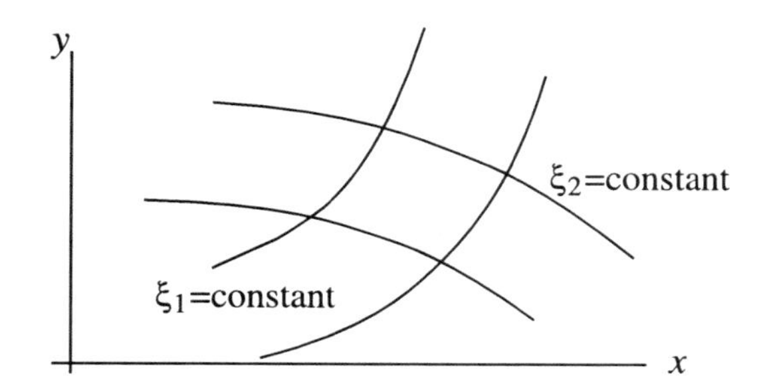

above Hamilton-Jacobi equation, i.e. the solutions of the system of ODE's:

$$\begin{aligned} \frac{\mathrm{d}x}{\mathrm{d}t} &= \frac{\partial H}{\partial p}, & \frac{\mathrm{d}p}{\mathrm{d}t} &= -\frac{\partial H}{\partial x} = 0, \\ \frac{\mathrm{d}z}{\mathrm{d}t} &= \frac{\partial H}{\partial q}, & \frac{\mathrm{d}q}{\mathrm{d}t} &= -\frac{\partial H}{\partial z} = 0 \end{aligned} \tag{2.95}$$

(see Sect. 9.2 for a brief summary of the concepts of characteristics). From the (2.95) it is easy to see that in the x, z, t-space, the characteristics are straight lines for constant t as in (2.91).

2.6 Wave Patterns

In Sect. 2.5, the Hamilton-Jacobi equation (2.94) for the wave fronts was derived from the equations in a rather complicated way. At first we gave an exact solution η of the linearised problem (1.32), (1.33) with $U = 0$, to which we later applied an asymptotic expansion, which resulted in a first-order partial differential equation. The result obtained is more or less similar to the characteristic equation for hyperbolic equations, although the wave fronts are by no means characteristic surfaces for the equations, which do not even have real characteristics.

In order to give a direct derivation we first define a *wave front* on the two-dimensional x, y-plane as a curve along which a transverse derivative of the solution φ of the equation considered is much larger than the tangential derivative. This means that, introducing new coordinates ξ_1 transverse to the wave fronts and ξ_2 along the wave fronts (Fig. 2.5), we must have that $\varphi_{\xi_1} \gg \varphi_{\xi_2}$, i.e. there should exist a constant $K \gg 1$ such that $\varphi_{\xi_1} \approx K\varphi_{\xi_2}$. Here ξ_1 and ξ_2 are supposed to be functions of x and y with derivatives of order unity with respect to K. We introduce a new coordinate $s = K\xi_1$ such that

$$\varphi_s = \frac{1}{K}\varphi_{\xi_1} = \mathrm{O}(1). \tag{2.96}$$

We now illustrate this procedure by considering a simpler equation than the equation of water waves, the Klein-Gordon equation in dimensionless form

$$\varphi_{xx} - \varphi_{tt} - a^2\varphi = 0, \tag{2.97}$$

where a is a constant. We first derive the Hamilton-Jacobi equations for the phase function to the methods used in Sect. 2.5 and will refer to it as an indirect method. For solutions of the form $Ae^{i(kx-\omega t)}$ we easily find the dispersion relation between k and ω,

$$\omega = \sqrt{a^2+k^2} \doteq H(k) \tag{2.98}$$

which gives the Hamilton-Jacobi equation from (2.94) with $\Psi = J$:

$$J_t + H(J_x) = 0, \tag{2.99}$$

where $J_x = k$. The characteristics of (2.99) are solutions of the equations

$$\begin{aligned} \frac{dx}{dt} &= \frac{\partial H}{\partial J_x} = \frac{k}{\sqrt{a^2+k^2}}, \\ \frac{dp}{dt} &= \frac{dJ_x}{dt} = -\frac{\partial H}{\partial x} = 0, \end{aligned} \tag{2.100}$$

thus the characteristics are straight lines of the form

$$x - \frac{k}{\sqrt{a^2+k^2}} t = \text{constant}, \tag{2.101}$$

corresponding to the group velocity $\frac{dH}{dk} = \frac{k}{\sqrt{a^2+k^2}}$.

Now let us examine the above problem by the direct method. Using (2.96), a straightforward calculation shows that

$$\begin{aligned} \varphi_{xx} &= K^2\varphi_{ss}\xi_{1x}^2 + K(2\varphi_{s\xi_2}\xi_{1x}\xi_{2x} + \xi_{1xx}\varphi_s) + \varphi_{\xi_2\xi_2}\xi_{2x}^2 + \varphi_{\xi_2}\xi_{2xx}, \\ \varphi_{tt} &= K^2\varphi_{ss}\xi_{1t}^2 + K(2\varphi_{s\xi_2}\xi_{1t}\xi_{2t} + \xi_{1tt}\varphi_s) + \varphi_{\xi_2\xi_2}\xi_{2t}^2 + \varphi_{\xi_2}\xi_{2tt}. \end{aligned} \tag{2.102}$$

Substituting into (2.97) gives

$$\begin{aligned} &K^2\varphi_{ss}(\xi_{1x}^2 - \xi_{1t}^2) + K\{\varphi_s(\xi_{1xx} - \xi_{1tt}) + 2\varphi_{s\xi_2}(\xi_{1x}\xi_{2x} - \xi_{1t}\xi_{2t})\} \\ &\quad + \varphi_{\xi_2\xi_2}(\xi_{2x}^2 - \xi_{2t}^2) + \varphi_{\xi_2}(\xi_{2t} - \xi_{2tt}) - a^2\varphi = 0. \end{aligned} \tag{2.103}$$

As $K \to \infty$, we obtain the characteristic equation for (2.97). This is obvious because a characteristic would be a line along which the second derivative may be discontinuous. Now, regarding the constant a as a large number with respect to some reference length and identifying K with a, we have the equation

$$\varphi_{ss}(\xi_{1x}^2 - \xi_{1t}^2) - \varphi = 0, \tag{2.104}$$

to the first order of approximation. If we want this equation to represent the motion along the wave fronts, we must put the term $(\xi_{1x}^2 - \xi_{1t}^2)$ equal to a constant which we choose to be -1, i.e.,

$$\xi_{1x}^2 - \xi_{1t}^2 = -1. \tag{2.105}$$

Clearly, this gives immediately the Hamilton-Jacobi equation, $\xi_{1t} = \sqrt{1 + \xi_{1x}^2}$, which reduces to (2.99) with ξ_1 replaced by $(-1/a)J$.

The same scheme can be applied to the problem of the moving singularity defined by the time independent form (1.32) and (1.33), i.e.:

$$\varphi_{xx} + \varphi_{yy} + \varphi_{zz} = 0,$$
$$U^2 \varphi_{xx} + g\varphi_y = 0, \quad \text{for } y = 0.$$

In terms of the dimensionless variables $\overline{x} = \frac{x}{L}$, $\overline{y} = \frac{y}{L}$ and $\overline{z} = \frac{z}{L}$, we have

$$\begin{aligned} &\varphi_{\overline{xx}} + \varphi_{\overline{yy}} + \varphi_{\overline{zz}} = 0, \\ &\varphi_{\overline{xx}} + \frac{gL}{U^2}\varphi_{\overline{y}} = 0, \quad \text{for } \overline{y} = 0. \end{aligned} \tag{2.106}$$

Here L denotes a proper reference length.

We are only interested in the *wave pattern*, hence in the lines of constant phase of η (which from (1.33) amounts to the same as for $\varphi_{\overline{x}}$ at $\overline{y} = 0$). We further remark that from the nature of (2.106) we know that the wave is only appreciable at the upper layer of the water. Hence we introduce the coordinates ξ_1 and ξ_2 in the x, z-plane, where the lines $\xi_1 =$ constant represent wave fronts, the derivative φ_{ξ_1} is large with respect to φ_{ξ_2} but the derivative $\varphi_{\overline{y}}$ must be of the same order of magnitude as φ_{ξ_1}. Therefore, we introduce a coordinate $s = K\xi_1$ and a coordinate $Y = K\overline{y}$ in terms of which we have

$$\varphi_{\overline{xx}} = K^2\varphi_{ss}\xi_{1\overline{x}}^2 + K(2\varphi_{s\xi_2}\xi_{1\overline{x}}\xi_{2\overline{x}} + \xi_{1\overline{xx}}\varphi_s) + \varphi_{\xi_2\xi_2}\xi_{2\overline{x}}^2 + \varphi_{\xi_2}\xi_{2\overline{xx}},$$
$$\varphi_{\overline{zz}} = K^2\varphi_{ss}\xi_{1\overline{z}}^2 + K(2\varphi_{s\xi_2}\xi_{1\overline{z}}\xi_{2\overline{z}} + \xi_{1\overline{zz}}\varphi_s) + \varphi_{\xi_2\xi_2}\xi_{2\overline{z}}^2 + \varphi_{\xi_2}\xi_{2\overline{zz}},$$

and

$$\varphi_{\overline{yy}} = K^2\varphi_{YY}.$$

From (2.106), we have then

$$K^2\varphi_{ss}(\xi_{1\overline{x}}^2 + \xi_{1\overline{z}}^2) + K^2\varphi_{YY} + \mathrm{O}(K) = 0, \tag{2.107}$$

together with the surface condition

$$K^2\varphi_{ss}\xi_{1\overline{x}}^2 + K\left(\frac{gL}{U^2}\right)\varphi_Y + \mathrm{O}(K) = 0, \quad \text{for } Y = 0. \tag{2.108}$$

This yields the first approximation

$$\begin{aligned} &\varphi_{ss}(\xi_{1\overline{x}}^2 + \xi_{1\overline{z}}^2) + \varphi_{YY} = 0, \\ &\varphi_{ss}\xi_{1\overline{x}}^2 + \varphi_Y = 0, \quad \text{for } Y = 0, \end{aligned} \tag{2.109}$$

where we identify K with $\frac{gL}{U^2}$.

Since $\xi_{1\bar{x}}^2 + \xi_{1\bar{z}}^2$ and $\xi_{1\bar{x}}^2$ are slowly varying variables, we introduce constants, α and β defined by

$$\alpha^2 = \xi_{1\bar{x}}^2 + \xi_{1\bar{z}}^2, \quad \text{and} \quad \beta = \xi_{1\bar{x}}^2.$$

This leads to the problem

$$\alpha^2\varphi_{ss} + \varphi_{YY} = 0,$$
$$\beta\varphi_{ss} + \varphi_Y = 0, \quad \text{for } Y = 0,$$

which has a solution

$$\varphi = \mathrm{e}^{\mathrm{i}s+.\alpha Y}.$$

This solution which goes to zero as $Y \to \infty$ ($\alpha > 0$) can satisfy the surface condition only if

$$\alpha = \beta$$

or

$$\xi_{1\bar{x}}^2 + \xi_{1\bar{z}}^2 = \xi_{1\bar{x}}^4. \tag{2.110}$$

It should be emphasised that these considerations are only valid to an order of magnitude of $1/K$. The present approach is a variation of the ray method in geometrical optics. Higher-order approximations can be derived in a similar manner.

The characteristic equations of the first-order partial differential equation (2.110) take the form

$$\dot{\bar{x}} = 4p^3 - 2p, \quad \dot{p} = 0,$$
$$\dot{\bar{z}} = -2q, \qquad \dot{q} = 0,$$
$$\dot{\xi_1} = (4p^4 - 2p^2 - 2q^2),$$

with $p = \xi_{1\bar{x}}$ and $q = \xi_{1\bar{z}}$, where the dot $\cdot$ notation denotes differentiation to some parameter, say, τ. Hence p and q are constants and we have the parametric equations for the rays,

$$\begin{aligned} \bar{x} &= 2p(2p^2 - 1)\tau, \\ \bar{z} &= -2q\tau, \\ \xi_1 &= (4p^4 - 2p^2 - 2q^2)\tau. \end{aligned} \tag{2.111}$$

From (2.110) we have

$$q = -p\sqrt{p^2 - 1}. \tag{2.112}$$

To eliminate τ from (2.111) by making use of (2.112), we finally obtain

$$\bar{x} = \xi_1 \frac{(2p^2 - 1)}{p^3}, \qquad \bar{z} = \xi_1 \frac{\sqrt{p^2 - 1}}{p^3},$$

which reduces to (2.88) if we set $p = -\frac{1}{\cos\theta}$. This shows that the curves $\xi_1 =$ constant are indeed the curves of constant phase.

2.7 Singularity in a Steady Current

2.7.1 Steady Singularity

As in the case of an oscillatory point source in still water it is useful to have the solution of a steady moving point source, or a point source in a steady current, available. For finite water depth the formulation becomes

$$\begin{aligned} &\varphi_{xx} + \varphi_{yy} + \varphi_{zz} = \delta(x - x_0, y - y_0, z - z_0), \\ &\varphi_y = 0 \qquad \text{at } y = -h, \\ &\upsilon\varphi_{xx} + \varphi_y = 0 \quad \text{at } y = 0, \end{aligned} \tag{2.113}$$

where we introduced the notation $\upsilon = \frac{U^2}{g}$. To obtain a physically valid solution we have to add a far-field condition, comparable with the radiation condition in the oscillatory case. Here the requirement becomes that in front of the disturbance no wavy pattern is observed. In the downstream region a wavy disturbance may be present. In the deep water case it is similar to the disturbance of the moving pressure point. It is also noticed that a solution of (2.113) can not be unique, because we always may add an arbitrary constant. We make use of this fact later. We follow the same procedure as described before (2.66) to solve (2.113),

$$\varphi(x, y, z) = -\frac{1}{4\pi r} - \frac{1}{4\pi \tilde{r}} + \psi(x, y, z), \tag{2.114}$$

where $\tilde{r}$ denotes the distance to reflected, with respect to the bottom, source point.

For $\psi(x, y, z)$ we obtain the formulation

$$\begin{aligned} &\psi_{xx} + \psi_{yy} + \psi_{zz} = 0, \\ &\psi_y = 0 \qquad\qquad \text{at} \quad y = -h, \\ &\psi_y + \upsilon\psi_{xx} = \frac{1}{4\pi}\left\{\frac{\partial}{\partial y}\left(\frac{1}{r} + \frac{1}{\tilde{r}}\right) + \upsilon\frac{\partial^2}{\partial x^2}\left(\frac{1}{r} + \frac{1}{\tilde{r}}\right)\right\} \quad \text{at } y = 0. \\ &\quad := r(x, z; x_0, y_0, z_0) \end{aligned} \tag{2.115}$$

If we apply the double Fourier transform to the function ψ,

$$\Psi(\alpha, y, \beta) = \int_{-\infty}^{\infty}\int_{-\infty}^{\infty} \mathrm{e}^{\mathrm{i}(\alpha x + \beta z)}\psi(x, y, z)\,\mathrm{d}x\,\mathrm{d}z, \tag{2.116}$$

we obtain the following ordinary differential equation and boundary conditions for Ψ:

$$\begin{aligned} &\Psi_{yy} - (\alpha^2 + \beta^2)\Psi = 0, && \\ &\Psi_y = 0 && \text{at } y = -h, \\ &\Psi_y - \upsilon\alpha^2\Psi = R(\alpha, \beta; x_0, y_0, z_0) && \text{at } y = 0. \end{aligned} \tag{2.117}$$

Application of (2.70) leads to the following expression for $R(\alpha, \beta; x_0, y_0, z_0)$,

$$R(\alpha, \beta; x_0, y_0, z_0) = -\frac{k + \upsilon\alpha^2}{k} \mathrm{e}^{-kh} \cosh(k(y_0 + h))\mathrm{e}^{\mathrm{i}(\alpha x_0 + \beta z_0)}, \tag{2.118}$$

where $k = \sqrt{\alpha^2 + \beta^2}$. The solution of (2.117) becomes

$$\Psi(\alpha, y, \beta) = -\frac{k + \upsilon\alpha^2}{k} \mathrm{e}^{-kh} \frac{\cosh(k(y + h))\cosh(k(y_0 + h))}{k\sinh(kh) - \upsilon\alpha^2\cosh(kh)} \mathrm{e}^{\mathrm{i}(\alpha x_0 + \beta z_0)}. \tag{2.119}$$

The inverse transform (2.73) of $\Psi(\alpha, y, \beta)$ becomes

$$\begin{aligned} \psi(x, y, z) = \frac{-1}{4\pi^2} \int_{-\infty}^{\infty} \int_{-\infty}^{\infty} &\mathrm{e}^{-kh} \cosh(k(y + h))\cosh(k(y_0 + h)) \\ &\cdot \frac{k + \upsilon\alpha^2}{k} \frac{\mathrm{e}^{-\mathrm{i}(\alpha(x - x_0) + \beta(z - z_0))}}{k\sinh(kh) - \upsilon\alpha^2\cosh(kh)} \,\mathrm{d}\alpha\,\mathrm{d}\beta \end{aligned} \tag{2.120}$$

and after the introduction of polar coordinates in the Fourier plane (2.52), (2.53)

$$\begin{aligned} \psi(x, y, z) = \frac{-1}{4\pi^2} \int_0^{\infty} \int_0^{2\pi} &\mathrm{e}^{-kh} \cosh(k(y + h))\cosh(k(y_0 + h)) \\ &\cdot \frac{(1 + k\upsilon\cos^2\vartheta)\mathrm{e}^{-\mathrm{i}k((x - x_0)\cos\vartheta + (z - z_0)\sin\vartheta)}}{\sinh(kh) - k\upsilon\cos^2\vartheta\cosh(kh)} \,\mathrm{d}k\,\mathrm{d}\vartheta. \end{aligned} \tag{2.121}$$

The integral is singular at $k = 0$. Therefor we make use of the fact that we may add a constant, with respect to x, y and z to the solution of (2.113). Hence a solution of (2.113) may be written as

$$\varphi(x, y, z) = -\frac{1}{4\pi r} - \frac{1}{4\pi\tilde{r}} + \tilde{\psi}(x, y, z) \tag{2.122}$$

with

$$\begin{aligned} \tilde{\psi}(x, y, z) = \frac{-1}{4\pi^2} \int_0^{\infty} \mathrm{d}k \int_0^{2\pi} \mathrm{d}\vartheta &\frac{\mathrm{e}^{-kh}}{\sinh(kh) - k\upsilon\cos^2\vartheta\cosh(kh)} \\ &\cdot \big\{\cosh(k(y + h))\cosh(k(y_0 + h))(1 + k\upsilon\cos^2\vartheta) \\ &\cdot \mathrm{e}^{-\mathrm{i}k((x - x_0)\cos\vartheta + (z - z_0)\sin\vartheta)} - 1\big\}. \end{aligned} \tag{2.123}$$

This solution does not fulfil the condition that upstream ($x \to -\infty$) no wavy disturbance may be present. To obey this condition a path of integration along the singularity on the real k-axis has to be chosen. Depending on the sign of $\cos\vartheta$ the choice will be different.

We notice that for $\cos\vartheta > 0$ and $x - x_0 > 0$ we may close the integral with respect to k in the fourth quadrant of the complex k-plane. For $\cos^2\vartheta < \frac{h}{\upsilon}$ we find a simple pole on the real k-axis. This means that to obtain a wavy contribution this singularity on the real axis must be inside the contour. We obtain a contribution of the pole plus an integral along the negative imaginary axis. This integral represents an exponentially decaying contribution. If however $x - x_0 < 0$ we close the integral in the first quadrant and we obtain a contribution of an integral along the positive imaginary axis only.

Next we consider $\cos\vartheta < 0$ and $x - x_0 > 0$ and we may close the integral in the first quadrant of the complex k-plane. We obtain a contribution of the singularity on the real axis if we chose the pole inside the contour. Again the integral along the imaginary axis is exponentially decaying. If $x - x_0 < 0$ we may close the contour in the fourth quadrant. This gives rise to a decaying contribution only.

We may reformulate the integral part of the solution by splitting up the integration with respect to ϑ into four parts of length $\pi/2$ and to combine the integral. In this way we obtain

$$\tilde{\psi}(x, y, z) = -\frac{1}{\pi^2}\int_0^\infty \mathrm{d}k \int_0^{\frac{\pi}{2}} \mathrm{d}\vartheta \frac{\mathrm{e}^{-kh}}{\sinh(kh) - k\upsilon\cos^2\vartheta\cosh(kh)}\{\cosh(k(y+h)) \cdot \cosh(k(y_0+h))(1 + k\upsilon\cos^2\vartheta)\cos((x-x_0) \cdot \cos\vartheta)\cos((z-z_0)\sin\vartheta) - 1\}. \tag{2.124}$$

In the Handbook of Physics [19], Wehausen gives further details of this expression.

To obtain an expression for the deep water case we let $h \to \infty$ in expression (2.123) and obtain.

$$\varphi(x, y, z) = -\frac{1}{4\pi r} + \frac{1}{4\pi\bar{r}} - \frac{1}{2\pi^2}\int_0^\infty \mathrm{d}k \int_0^\pi \mathrm{d}\vartheta \frac{1}{1 - k\upsilon\cos^2\vartheta} \cdot \mathrm{e}^{k((y+y_0) - \mathrm{i}(x-x_0)\cos\vartheta)}\cos(k(z-z_0)\sin\vartheta) \tag{2.125}$$

where we used (2.70) to obtain the contribution of a singularity at the point $(x_0, -y_0, z_0)$, hence $\bar{r}$ is defined as $\sqrt{(x-x_0)^2 + (y+y_0)^2 + (z-z_0)^2}$. The contour in the k-plane has to be chosen as before. For $\cos\vartheta > 0$ the contour passes the singularity in the upper plane, while for $\cos\vartheta < 0$ the contour passes the singularity in the lower plane. The contribution of the pole gives a far-field pattern comparable with the moving pressure point wave field described in Sect. 2.4.

2.7.2 Oscillating Singularity

The boundary value problem for the disturbance of a steady flow is described in (1.32). We consider a harmonic point source and assume that the potential function can be written as

$$\phi(x, y, z, t) = \mathrm{e}^{\mathrm{i}\omega t}\hat{\phi}(x, y, z).$$

The boundary value problem for the disturbance of a point source in (x_0, y_0, z_0) becomes

$$\varphi_{xx} + \varphi_{yy} + \varphi_{zz} = \delta(x - x_0, y - y_0, z - z_0)\mathrm{e}^{\mathrm{i}\omega t}. \tag{2.126}$$

If we introduce $\varphi(x, y, z, t) = \hat{\varphi}(x, y, z)\mathrm{e}^{\mathrm{i}\omega t}$, the boundary value problem to be solved becomes

$$\begin{aligned} &\hat{\varphi}_{xx} + \hat{\varphi}_{yy} + \hat{\varphi}_{zz} = \delta(x - x_0, y - y_0, z - z_0), \\ &\hat{\varphi}_y = 0 \qquad \text{at } y = -h, \\ &\upsilon\hat{\phi}_{xx} + 2\mathrm{i}\tau\hat{\varphi}_x - \nu\hat{\varphi} + \hat{\varphi}_y = 0 \quad \text{at } y = 0, \end{aligned} \tag{2.127}$$

where we introduced the parameters $\nu = \omega^2/g$, $\upsilon = U^2/g$,and $\tau = (\omega U)/g$; notice that $\tau^2 = \nu\upsilon$.

$$\hat{\varphi}(x, y, z) = -\frac{1}{4\pi r} - \frac{1}{4\pi\tilde{r}} + \psi(x, y, z). \tag{2.128}$$

Introduction of the double Fourier transform leads to the following ordinary differential equation for the transform of ψ,

$$\begin{aligned} &\Psi_{yy} - (\alpha^2 + \beta^2)\Psi = 0, \\ &\Psi_y = 0 \qquad \text{at } y = -h, \\ &\Psi_y - (\upsilon\alpha^2 + 2\tau\alpha + \nu)\Psi = S(\alpha, \beta; x_0, y_0, z_0) \quad \text{at } y = 0. \end{aligned} \tag{2.129}$$

Application of (2.70) leads to the following expression for $S(\alpha, \beta; x_0, y_0, z_0)$,

$$S(\alpha, \beta; x_0, y_0, z_0) = -\frac{k + \upsilon\alpha^2 + 2\tau\alpha + \nu}{k}\mathrm{e}^{-kh}\cosh(k(y_0 + h))\mathrm{e}^{\mathrm{i}(\alpha x_0 + \beta z_0)}, \tag{2.130}$$

where $k = \sqrt{\alpha^2 + \beta^2}$. The solution of (2.129) becomes

$$\begin{aligned} \Psi(\alpha, y, \beta) = &-\frac{k + \upsilon\alpha^2 + 2\tau\alpha + \nu}{k}\mathrm{e}^{-kh} \\ &\cdot \frac{\cosh(k(y + h))\cosh(k(y_0 + h))}{k\sinh(kh) - (\upsilon\alpha^2 + 2\tau\alpha + \nu)\cosh(kh)}\mathrm{e}^{\mathrm{i}(\alpha x_0 + \beta z_0)}. \end{aligned} \tag{2.131}$$

The inverse transform of $\Psi(\alpha, y, \beta)$ gives the solution of (2.127). The choice of the path of integration will be elucidated in the deep water case. Hence we consider the limit $h \to \infty$. We rewrite expression (2.131) in the form

$$\Psi(\alpha, y, \beta) = \left(\frac{1}{2} - \mathscr{L}(k, \alpha)\right) \frac{e^{k(y+y_0)} e^{i(\alpha x_0 + \beta z_0)}}{k}. \tag{2.132}$$

The function $\mathscr{L}(k, \alpha)$ becomes, in polar coordinates in the Fourier plane,

$$\lim_{h \to \infty} \mathscr{L}(k, \theta) = \frac{k}{k - (\upsilon\alpha^2 + 2\tau\alpha + \nu)} = \frac{gk}{gk - (\omega + kU\cos\theta)^2}. \tag{2.133}$$

Finally we obtain an expression for $\hat{\varphi}(x, y, z)$ where we still have to choose the proper path of integration in the complex k-plane

$$\varphi(x, y, z) = -\frac{1}{4\pi r} + \frac{1}{4\pi \bar{r}} - \frac{1}{2\pi^2} \int_0^\infty \int_0^\pi \frac{gk}{gk - (\omega + kU\cos\theta)^2} \cdot e^{k((y+y_0) - i(x-x_0)\cos\vartheta)} \cos(k(z - z_0)\sin\vartheta)\, dk\, d\vartheta. \tag{2.134}$$

We will investigate the zeros of the denominator. The quadratic equation has two zeros,

$$gk^{\pm} = \frac{1 - 2\tau\cos\vartheta \pm \sqrt{1 - 4\tau\cos\vartheta}}{2\tau^2\cos^2\vartheta}\omega^2. \tag{2.135}$$

First of all we notice that, for values of ϑ for which we have

$$1 - 4\tau\cos\vartheta < 0, \tag{2.136}$$

we find no singularities of the integrand along the real k-axis. Hence for $\tau > 1/4$ we find a ϑ interval where the k-integral is regular for $0 \le \vartheta < \gamma$ with $\cos\gamma = 1/(4\tau)$. For $\tau < 1/4$ we have $\gamma = 0$. Next, to determine the contour of integration when two poles lie on the positive real axis we have to consider the condition in the far field. It is easy to show that for $\vartheta > \gamma$ both roots of the quadratic equation are situated on the positive real axis of the complex k-plane. It is convenient to consider the poles for small values of U and ω successively. In both cases τ becomes small, so we consider the two poles for $\tau \to 0$. We find

$$\lim_{\tau \to 0} k^- = \frac{\omega^2}{g} \quad \text{and} \quad \lim_{\tau \to 0} k^+ = \frac{g}{U^2\cos^2\vartheta}. \tag{2.137}$$

In the oscillatory case without current we have seen that the contour passes the pole in the first quadrant of the complex k-plane, for all values of ϑ. Actually we could carry out the ϑ integral in that case. Hence, we conclude that this is also the case for the singularity in k^-.

In the case of a steady source in a current we have seen that we have to consider the sign of $\cos\vartheta$ because in the downstream direction the far field shows a wavy character. Hence for $\cos\vartheta > 0$ the contour of integration passes k^+ in the first

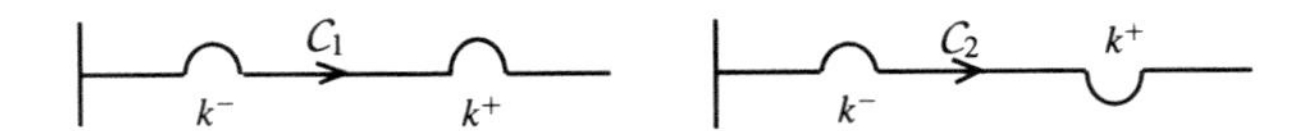

Fig. 2.6 Lines of integration

quadrant, while for $\cos\vartheta < 0$ the contour passes k^+ in the fourth quadrant of the complex k-plane (see Fig. 2.6). Finally the solution can be written as

$$\varphi(x,y,z) = -\frac{1}{4\pi r} + \frac{1}{4\pi \bar{r}} - \frac{1}{2\pi^2}\left\{\int_0^\infty \int_0^\gamma + \int_{\mathscr{C}_1}\int_\gamma^{\frac{\pi}{2}} + \int_{\mathscr{C}_2}\int_{\frac{\pi}{2}}^{\pi}\right\}$$
$$\cdot \frac{gk\mathrm{e}^{k((y+y_0)-\mathrm{i}(x-x_0)\cos\vartheta)}}{gk-(\omega+kU\cos\theta)^2}\cos(k(z-z_0)\sin\vartheta)\,\mathrm{d}k\,\mathrm{d}\vartheta. \qquad (2.138)$$

Chapter 3
Boundary Integral Formulation and Ship Motions

In the field of ship hydrodynamics the application of integral equations to solve the linear ship motion problem is widely used. For linear problems harmonic in time there are different ways to formulate an integral equation. A popular formulation, described in this chapter, is the one in the frequency domain. A less frequently used approach is a formulation in the time domain. The advantage of the latter approach is that the source function is rather simple and that it can be extended to the non linear case with some minor effort. For the steady ship wave problem a non-linear approach is the currently most used. In this chapter we give an introduction to these formulations. These methods consist of a description by means of regular Fredholm integral equations. In this chapter we present the method based on Green's theorem. To outline the method and the difficulties encountered we treat a simpler case, namely the diffraction of an acoustic wave by a smooth object. Hence we first consider the Helmholz equation for the scattering of acoustic waves, instead of the free surface problem.

A short introduction is given to the linear ship motion problem. The structure of the equations of motion is explained by the treatment of an object that is free to move in one degree of freedom, namely the vertical heave motion. The coefficients equations of motion of the general problem, with six degrees of freedom, can be determined by means of the numerical solution of the integral equations for the potential. In the same way the exciting forces may be computed. For the zero forward speed case this approach is generally used. Extension to the forward speed case is possible if the steady flow field around the ship does not differ much from the uniform velocity. Here we describe a way to apply a slow speed approximation.

3.1 Scattering of Acoustic Waves

The equation for a time dependent acoustic disturbance $\bar{u}(x, y, z, t)$ is the well-known wave equation

$$\bar{u}_{xx} + \bar{u}_{yy} + \bar{u}_{zz} = \frac{1}{c^2}\bar{u}_{tt}, \tag{3.1}$$

A.J. Hermans, *Water Waves and Ship Hydrodynamics*,
DOI 10.1007/978-94-007-0096-3_3, © Springer Science+Business Media B.V. 2011

where c is the specific velocity of sound in the medium. We look for harmonic wave solutions of this equation. Hence we look for solutions of the form

$$\bar{u}(x, y, z, t) = u(x, y, z)e^{i\omega t}. \tag{3.2}$$

The function u now becomes a solution of the Helmholz equation

$$u_{xx} + u_{yy} + u_{zz} + \frac{\omega^2}{c^2}u = \Delta u + k^2 u = 0 \tag{3.3}$$

where Δ is the three-dimensional Laplace operator and $k = \frac{\omega}{c}$ is the wave number. The acoustic field due to a point source is a solution of

$$\Delta\tilde{u} + k^2\tilde{u} = \delta(x - \xi, y - \eta, z - \zeta). \tag{3.4}$$

The solution can be constructed by either introduction of spherical coordinates or in analogy with the method used in Sect. 2.3 by means of a double Fourier transform performed for instance in the x and z coordinates. We follow the former procedure. The solution of (3.4) only depends on the distance

$$r = \sqrt{(x - \xi)^2 + (y - \eta)^2 + (z - \zeta)^2} = |\boldsymbol{x} - \boldsymbol{\xi}|$$

to the source. We obtain the following ordinary differential equation for $\tilde{u}(R)$,

$$\tilde{u}_{rr} + \frac{2}{r}\tilde{u}_r + k^2\tilde{u} = 0 \quad \text{for } r \neq 0. \tag{3.5}$$

Hence the singular solution of (3.4) becomes a linear combination of $\tilde{u}^{\pm} = \frac{e^{\pm ik|\boldsymbol{x}-\boldsymbol{\xi}|}}{|\boldsymbol{x}-\boldsymbol{\xi}|}$. The solution of the source problem must consist of outgoing waves only, hence we choose

$$\tilde{u} = -\frac{1}{4\pi}\frac{e^{-ik|\boldsymbol{x}-\boldsymbol{\xi}|}}{|\boldsymbol{x} - \boldsymbol{\xi}|}. \tag{3.6}$$

The choice of the factor $-\frac{1}{4\pi}$ is easily checked by taking the limit case for $k \to 0$. The result reduces to the singular solution of the Laplace equation where we require a unit outgoing flux through a sphere around the source.

We shall construct some possible boundary integral formulation for the scattering problem. We consider a plane incident wave scattered by a smooth penetrable or impenetrable convex object. The general boundary condition on the object is $Au + B\frac{\partial u}{\partial n} = 0$, where n is along the normal at the scattering surface $\mathscr{S}$, see Fig. 3.1. In analogy with the ship wave problem we take as boundary condition $\frac{\partial u}{\partial n} = 0$. To obtain an integral equation along the surface we first consider a point (x, y, z) in the domain $\mathscr{D}^-$, inside the object and apply Green's theorem to the functions u and $\tilde{u}$. In the case of exterior scattering this is an artificial problem. We do so because it will give us insight into the behaviour the mathematical formulation of the exterior

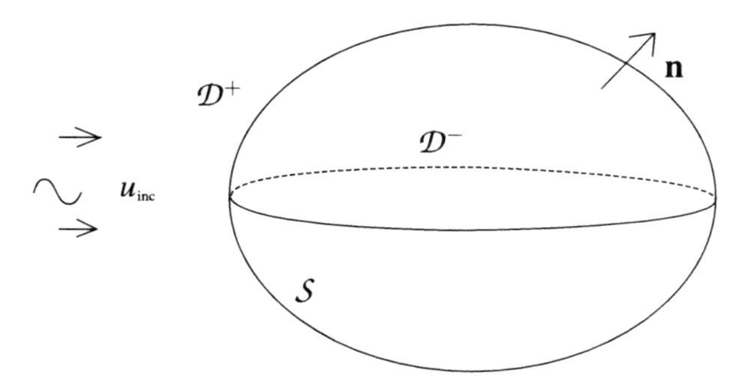

Fig. 3.1 Scattering object

physical scattering problem. We denote the field in $\mathscr{D}^-$ by u^-. The Green's theorem can be written in the form

$$\iiint\limits_{\mathscr{D}^-} (u\Delta\tilde{u} - \tilde{u}\Delta u)\,\mathrm{d}V = \iint\limits_{\mathscr{S}} \left(u\frac{\partial\tilde{u}}{\partial n} - \tilde{u}\frac{\partial u}{\partial n}\right)\mathrm{d}s, \tag{3.7}$$

where $\mathbf{n}$ is the outward normal to the surface of the object. Notice that $\tilde{u}$ is a solution of (3.4), hence we have to distinguish three positions of the point $\boldsymbol{x} = (x, y, z)$. We find

$$\begin{aligned}
u^-(\boldsymbol{x}) &= \iint\limits_{\mathscr{S}} \left(u^-(\boldsymbol{\xi})\frac{\partial\tilde{u}(\boldsymbol{x},\boldsymbol{\xi})}{\partial n_\xi} - \tilde{u}(\boldsymbol{x},\boldsymbol{\xi})\frac{\partial u^-(\boldsymbol{\xi})}{\partial n_\xi}\right)\mathrm{d}s_\xi \quad \text{for } \boldsymbol{x}\in\mathscr{D}^-,\\
\frac{u^-(\boldsymbol{x})}{2} &= \iint\limits_{\mathscr{S}} \left(u^-(\boldsymbol{\xi})\frac{\partial\tilde{u}(\boldsymbol{x},\boldsymbol{\xi})}{\partial n_\xi} - \tilde{u}(\boldsymbol{x},\boldsymbol{\xi})\frac{\partial u^-(\boldsymbol{\xi})}{\partial n_\xi}\right)\mathrm{d}s_\xi \quad \text{for } \boldsymbol{x}\in\mathscr{S},\\
0 &= \iint\limits_{\mathscr{S}} \left(u^-(\boldsymbol{\xi})\frac{\partial\tilde{u}(\boldsymbol{x},\boldsymbol{\xi})}{\partial n_\xi} - \tilde{u}(\boldsymbol{x},\boldsymbol{\xi})\frac{\partial u^-(\boldsymbol{\xi})}{\partial n_\xi}\right)\mathrm{d}s_\xi \quad \text{for } \boldsymbol{x}\in\mathscr{D}^+.
\end{aligned} \tag{3.8}$$

The field in the external region $\mathscr{D}^+$ can be written as $u_{inc} + u^+$. Application of the Green's theorem in the external field yields

$$\begin{aligned}
0 &= \iint\limits_{\mathscr{S}} \left(u^+(\boldsymbol{\xi})\frac{\partial\tilde{u}(\boldsymbol{x},\boldsymbol{\xi})}{\partial n_\xi} - \tilde{u}(\boldsymbol{x},\boldsymbol{\xi})\frac{\partial u^+(\boldsymbol{\xi})}{\partial n_\xi}\right)\mathrm{d}s_\xi + \mathscr{I}_\infty \quad \text{for } \boldsymbol{x}\in\mathscr{D}^-,\\
-\frac{u^+(\boldsymbol{x})}{2} &= \iint\limits_{\mathscr{S}} \left(u^+(\boldsymbol{\xi})\frac{\partial\tilde{u}(\boldsymbol{x},\boldsymbol{\xi})}{\partial n_\xi} - \tilde{u}(\boldsymbol{x},\boldsymbol{\xi})\frac{\partial u^+(\boldsymbol{\xi})}{\partial n_\xi}\right)\mathrm{d}s_\xi + \mathscr{I}_\infty \quad \text{for } \boldsymbol{x}\in\mathscr{S},\\
-u^+(\boldsymbol{x}) &= \iint\limits_{\mathscr{S}} \left(u^+(\boldsymbol{\xi})\frac{\partial\tilde{u}(\boldsymbol{x},\boldsymbol{\xi})}{\partial n_\xi} - \tilde{u}(\boldsymbol{x},\boldsymbol{\xi})\frac{\partial u^+(\boldsymbol{\xi})}{\partial n_\xi}\right)\mathrm{d}s_\xi + \mathscr{I}_\infty \quad \text{for } \boldsymbol{x}\in\mathscr{D}^+.
\end{aligned} \tag{3.9}$$

The minus sign on the left-hand side is due to the fact that the normal is directed towards the interior of the domain $\mathscr{D}^+$. The contribution of a surface in the far field is denoted by $\mathscr{I}_\infty$. We may take a sphere, $\mathscr{S}_\infty$, with centre at (x_0, y_0, z_0) and radius r. The far-field integral can then be written as

$$\mathscr{I}_\infty = -\iint\limits_{\mathscr{S}_\infty} \left(u^+ \frac{\partial \tilde{u}}{\partial r} - \tilde{u} \frac{\partial u^+}{\partial r} \right) \mathrm{d}s$$

$$= \lim_{r \to \infty} \frac{1}{4\pi} \int\limits_{-\frac{\pi}{2}}^{\frac{\pi}{2}} \int\limits_{0}^{2\pi} \left(u^+ \frac{\partial \frac{\mathrm{e}^{-ikr}}{r}}{\partial r} - \frac{\mathrm{e}^{-ikr}}{r} \frac{\partial u^+}{\partial r} \right) r^2 \sin(\theta)\, \mathrm{d}\theta\, \mathrm{d}\phi. \quad (3.10)$$

The appropriate far-field conditions are

$$\lim_{r \to \infty} u^+ = 0 \quad \text{and} \quad \lim_{r \to \infty} r \left(iku^+ + \frac{\partial u^+}{\partial r} \right) = 0. \quad (3.11)$$

It is easy to show that r in this formulation can be taken as the distance to the origin of the coordinate system. The latter condition is the well-known *Sommerfeld* radiation condition. Only outgoing waves are permitted in the scattered far field. We notice that if we had taken as time dependency $\mathrm{e}^{-\mathrm{i}\omega t}$, the sign in the radiation condition changes. So from now on we take $\mathscr{I}_\infty = 0$ in formulation (3.9).

There are several ways to obtain an integral equation for the exterior Neumann problem. Hence the boundary condition for the function u^+ becomes $\frac{\partial u^+}{\partial n} = -\frac{\partial u_{inc}}{\partial n}$ on $\mathscr{S}$.

3.1.1 Direct Method

The second relation in (3.9) gives an integral equation for the function u^+ on $\mathscr{S}$. After solving this equation numerically, the function u^+ in the domain $\mathscr{D}^+$ is then given by the third relation in (3.9). In a slightly different, commonly used, notation the integral equation becomes

$$\frac{1}{2} u^+(\mathbf{x}) = -\iint\limits_{\mathscr{S}} \left(u^+(\boldsymbol{\xi}) \frac{\partial G(\boldsymbol{x}, \boldsymbol{\xi})}{\partial n_{\xi}} + G(\boldsymbol{x}, \boldsymbol{\xi}) \frac{\partial u_{inc}(\boldsymbol{\xi})}{\partial n_{\xi}} \right) \mathrm{d}s_{\xi}$$

$$\text{for } \boldsymbol{x} \in \mathscr{S}, \quad (3.12)$$

where

$$G(\boldsymbol{x}, \boldsymbol{\xi}) = \tilde{u}(\boldsymbol{x}, \boldsymbol{\xi}) = -\frac{1}{4\pi} \frac{\mathrm{e}^{-ikr}}{r}$$

with $r = |\boldsymbol{x} - \boldsymbol{\xi}|$.

3.1.2 Source Distribution

Let us combine (3.8) and (3.9). If we subtract the two expressions in the domain $\mathscr{D}^+$ we obtain the following expression for u^+

$$-u^+(\boldsymbol{x}) = \iint_{\mathscr{S}} \left((u^+(\boldsymbol{\xi}) - u^-(\boldsymbol{\xi}))\frac{\partial G(\boldsymbol{x},\boldsymbol{\xi})}{\partial n_{\xi}} - G(\boldsymbol{x},\boldsymbol{\xi})\frac{\partial (u^+(\boldsymbol{\xi}) - u^-(\boldsymbol{\xi}))}{\partial n_{\xi}} \right) \mathrm{d}s_{\xi}. \tag{3.13}$$

The inner problem is an artificial one, so we may choose its boundary condition on $\mathscr{S}$. We take

$$u^+(\boldsymbol{\xi}) = u^-(\boldsymbol{\xi}) \quad \text{for } \boldsymbol{\xi} \in \mathscr{S}$$

and define

$$\sigma(\boldsymbol{\xi}) = \frac{\partial (u^+(\boldsymbol{\xi}) - u^-(\boldsymbol{\xi}))}{\partial n_{\xi}} \quad \text{for } \boldsymbol{\xi} \in \mathscr{S}. \tag{3.14}$$

The expression for u^+ now becomes

$$u^+(\boldsymbol{x}) = \iint_{\mathscr{S}} \sigma(\boldsymbol{\xi})G(\boldsymbol{x},\boldsymbol{\xi})\,\mathrm{d}s_{\xi}. \tag{3.15}$$

This expression can be interpreted as a distribution of sources, over $\mathscr{S}$, with strength $\sigma(\boldsymbol{\xi})$ and influence function $G(\boldsymbol{x},\boldsymbol{\xi})$. For the Neumann problem we obtain an integral equation by differentiating (3.15) with respect to the normal direction at the surface. We obtain

$$\frac{\partial u^+(\boldsymbol{x})}{\partial n_{\boldsymbol{x}}} = -\frac{\partial u_{inc}(\boldsymbol{x})}{\partial n_{\boldsymbol{x}}} = \frac{1}{2}\sigma(\boldsymbol{x}) + \iint_{\mathscr{S}} \sigma(\boldsymbol{\xi})\frac{\partial G(\boldsymbol{x},\boldsymbol{\xi})}{\partial n_{\boldsymbol{x}}}\,\mathrm{d}s_{\xi}. \tag{3.16}$$

The question now arises whether this integral equation has a unique solution, when the exterior problem has a unique solution. The answer to this question is that for certain frequencies the integral equation has a non-unique solution. This is easy to show by studying the artificial interior formulation for a homogeneous Dirichlet problem, i.e., $u^- = 0$ for $\boldsymbol{x} \in \mathscr{S}$. In this case we may formulate an integral equation for the normal derivative u_n^- at $\mathscr{S}$. The first expression in (3.8) reads

$$u^-(\boldsymbol{x}) = -\iint_{\mathscr{S}} G(\boldsymbol{x},\boldsymbol{\xi})\frac{\partial u^-(\boldsymbol{\xi})}{\partial n_{\xi}}\,\mathrm{d}s_{\xi}.$$

Differentiation of this expression in the direction of the normal yields

$$\frac{\partial u^-(\boldsymbol{x})}{\partial n_{\boldsymbol{x}}} = -\iint_{\mathscr{S}} \frac{\partial G(\boldsymbol{x},\boldsymbol{\xi})}{\partial n_{\boldsymbol{x}}}\frac{\partial u^-(\boldsymbol{\xi})}{\partial n_{\xi}}\,\mathrm{d}s_{\xi} \quad \text{for } \boldsymbol{x} \in \mathscr{D}^-.$$

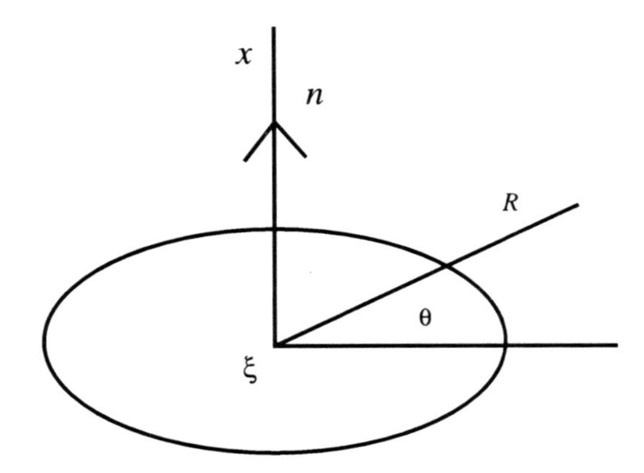

Fig. 3.2 Local approximation

If we let $\boldsymbol{x}$ approach $\mathscr{S}$ from the inside region $\mathscr{D}^-$ we have to evaluate the integral carefully near the singular point. We consider the integral over the surface of a small circle, with radius ρ, around the projection of the singular point, on $\mathscr{S}$. We may consider the surface locally flat, see Fig. 3.2.

The integral may be approximated, for small values of the distance ε to the surface, by

$$\frac{1}{4\pi}\frac{\partial u^-(\boldsymbol{x})}{\partial n_{\boldsymbol{x}}}\int_0^{\rho}\int_0^{2\pi}\frac{\varepsilon}{(R^2+\varepsilon^2)^{3/2}}R\,\mathrm{d}R\,\mathrm{d}\theta=\frac{1}{2}\frac{\partial u^-(\boldsymbol{x})}{\partial n_{\boldsymbol{x}}}+\mathrm{O}(\varepsilon). \tag{3.17}$$

Hence we obtain a relation for u_n^- at the surface:

$$\frac{1}{2}\frac{\partial u^-(\boldsymbol{x})}{\partial n_{\boldsymbol{x}}}+\iint_{\mathscr{S}}\frac{\partial G(\boldsymbol{x},\boldsymbol{\xi})}{\partial n_{\boldsymbol{x}}}\frac{\partial u^-(\boldsymbol{\xi})}{\partial n_{\boldsymbol{\xi}}}\,\mathrm{d}s_{\boldsymbol{\xi}}=0 \quad \text{for } \boldsymbol{x}\in\mathscr{S}. \tag{3.18}$$

The integral should be interpreted as the principle value integral over the surface $\mathscr{S}$. It is well known that, at certain frequencies, the internal homogeneous Dirichlet problem exhibits resonance. Hence (3.18) has non-zero solutions at these frequencies. This means that the equation for the source function (3.16) has non-unique solutions at the same frequencies, while in the exterior region no resonance phenomenon occurs. The attempt to find a solution of the exterior Neumann problem by means of a source distribution breaks down at these *irregular* frequencies.

3.2 Scattering of Free Surface Waves

For offshore engineering one is interested in the forces and moments acting on fixed or floating structures. To compute these forces one must take into account the effect of the scattered field. Hence, we first consider the scattering of free surface waves due to a fixed floating object. The total wave field results in a pressure distribution

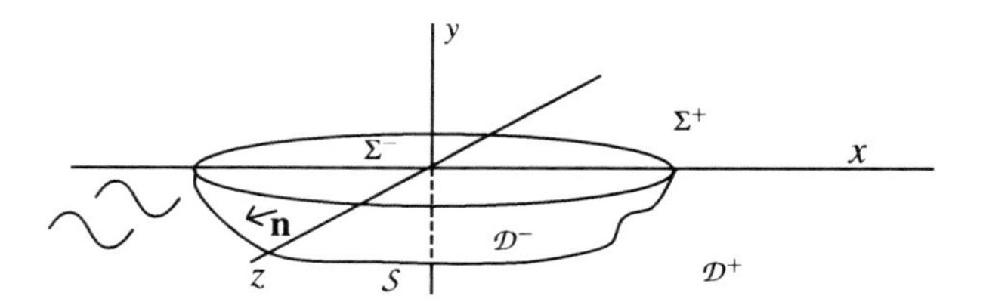

Fig. 3.3 Object in waves

along the object. Hence the forces and moments may be computed. If the body is free to move in its six degrees of freedom it excites waves as well, resulting in the so-called *reaction* forces and moments. The forces and moments can be split up in a part in phase with the motion and a part out of phase with the motion. The first one is generally associated with the so-called *added mass* and the second with the *wave damping*.

3.2.1 Fixed Object

We first consider the scattering of waves at finite water depth by a fixed object, see Fig. 3.3. The potential in the outside domain $\mathscr{D}^+$ is written as

$$\varphi(x,y,z,t)=\hat{\varphi}(x,y,z)\mathrm{e}^{\mathrm{i}\omega t}=(\phi_{inc}(x,y,z)+\phi(x,y,z))\mathrm{e}^{\mathrm{i}\omega t}. \tag{3.19}$$

This potential function $\phi(x,y,z)$ has to obey the linearised free surface condition at Σ^+. In the inner domain $\mathscr{D}^-$ we define an auxiliary potential function $\phi^-(x,y,z)$. In principle one is free to choose a condition at the *free* surface Σ^- inside the body. Later we will make use of this fact to avoid the irregular frequencies. We consider the water depth to be finite. We apply Green's theorem in the Domain $\mathscr{D}^+$:

$$\iiint\limits_{\mathscr{D}^+}(\phi\Delta G-G\Delta\phi)\,\mathrm{d}V=-\iint\limits_{\overline{\mathscr{S}}}\left(\phi(\boldsymbol{\xi})\frac{\partial G(\boldsymbol{x},\boldsymbol{\xi})}{\partial n_{\xi}}-G(\boldsymbol{x},\boldsymbol{\xi})\frac{\partial\phi(\boldsymbol{\xi})}{\partial n_{\xi}}\right)\mathrm{d}s_{\xi}, \tag{3.20}$$

where $\mathbf{n}$ is the normal vector pointing into the domain $\mathscr{D}^+$. $\overline{\mathscr{S}}$ is the total bounding surface of the fluid domain consisting of $\mathscr{S}$, the bottom $\mathscr{B}$, the free surface Σ^+ and a closing surface $\mathscr{S}^\infty$ in the far field. For the source function, which is often called the Green's function, we choose (2.65) or (2.75) for the finite water depth case and (2.76) for the deep water case. Doing so the integral over the bounding surface reduces to an integral over the body surface $\mathscr{S}$ and the surface at infinity $\mathscr{S}^\infty$. If we make use of the appropriate radiation condition for the scattered waves,

$$\lim_{R\to\infty}\phi=0 \quad \text{and} \quad \lim_{R\to\infty}\sqrt{R}\left(\mathrm{i}k\phi+\frac{\partial\phi}{\partial R}\right)=0 \tag{3.21}$$

where R is the horizontal distance to the origin of the coordinate system the integral over $\mathscr{S}^\infty$ vanishes as well and $\overline{\mathscr{S}}$ reduces to $\mathscr{S}$. If we choose the source function as mentioned we obtain

$$
\begin{aligned}
0 &= -\iint\limits_{\mathscr{S}} \left(\phi(\boldsymbol{\xi})\frac{\partial G(\boldsymbol{x},\boldsymbol{\xi})}{\partial n_\xi} - G(\boldsymbol{x},\boldsymbol{\xi})\frac{\partial \phi(\boldsymbol{\xi})}{\partial n_\xi}\right) \mathrm{d}s_\xi \qquad \text{for } \boldsymbol{x} \in \mathscr{D}^-,\\
\frac{\phi(\boldsymbol{x})}{2} &= -\iint\limits_{\mathscr{S}} \left(\phi(\boldsymbol{\xi})\frac{\partial G(\boldsymbol{x},\boldsymbol{\xi})}{\partial n_\xi} - G(\boldsymbol{x},\boldsymbol{\xi})\frac{\partial \phi(\boldsymbol{\xi})}{\partial n_\xi}\right) \mathrm{d}s_\xi \qquad \text{for } \boldsymbol{x} \in \mathscr{S},\\
\phi(\boldsymbol{x}) &= -\iint\limits_{\mathscr{S}} \left(\phi(\boldsymbol{\xi})\frac{\partial G(\boldsymbol{x},\boldsymbol{\xi})}{\partial n_\xi} - G(\boldsymbol{x},\boldsymbol{\xi})\frac{\partial \phi(\boldsymbol{\xi})}{\partial n_\xi}\right) \mathrm{d}s_\xi \qquad \text{for } \boldsymbol{x} \in \mathscr{D}^+.
\end{aligned}
\tag{3.22}
$$

Application of the Green theorem in the domain inside the floating body leads to a relation for the auxiliary potential ϕ^-:

$$
\begin{aligned}
\phi^-(\boldsymbol{x}) &= \iint\limits_{\mathscr{S}} \left(\phi^-(\boldsymbol{\xi})\frac{\partial G(\boldsymbol{x},\boldsymbol{\xi})}{\partial n_\xi} - G(\boldsymbol{x},\boldsymbol{\xi})\frac{\partial \phi^-(\boldsymbol{\xi})}{\partial n_\xi}\right) \mathrm{d}s_\xi + \mathscr{I}^\Sigma \qquad \text{for } \boldsymbol{x} \in \mathscr{D}^-,\\
\frac{\phi^-(\boldsymbol{x})}{2} &= \iint\limits_{\mathscr{S}} \left(\phi^-(\boldsymbol{\xi})\frac{\partial G(\boldsymbol{x},\boldsymbol{\xi})}{\partial n_\xi} - G(\boldsymbol{x},\boldsymbol{\xi})\frac{\partial \phi^-(\boldsymbol{\xi})}{\partial n_\xi}\right) \mathrm{d}s_\xi + \mathscr{I}^\Sigma \qquad \text{for } \boldsymbol{x} \in \mathscr{S},\\
0 &= \iint\limits_{\mathscr{S}} \left(\phi^-(\boldsymbol{\xi})\frac{\partial G(\boldsymbol{x},\boldsymbol{\xi})}{\partial n_\xi} - G(\boldsymbol{x},\boldsymbol{\xi})\frac{\partial \phi^-(\boldsymbol{\xi})}{\partial n_\xi}\right) \mathrm{d}s_\xi + \mathscr{I}^\Sigma \qquad \text{for } \boldsymbol{x} \in \mathscr{D}^+,
\end{aligned}
\tag{3.23}
$$

with

$$
\mathscr{I}^\Sigma = \iint\limits_{\Sigma^-} \left(\phi^-(\boldsymbol{\xi})\frac{\partial G(\boldsymbol{x},\boldsymbol{\xi})}{\partial \eta} - G(\boldsymbol{x},\boldsymbol{\xi})\frac{\partial \phi^-(\boldsymbol{\xi})}{\partial \eta}\right) \mathrm{d}\xi\,\mathrm{d}\zeta. \tag{3.24}
$$

Making use of the free surface condition (2.35) for the source function we obtain

$$
\mathscr{I}^\Sigma = \iint\limits_{\Sigma^-} \left(\frac{\omega^2}{g}\phi^-(\xi,0,\zeta) - \frac{\partial \phi^-(\xi,0,\zeta)}{\partial \eta}\right) G(\boldsymbol{x},\xi,0,\zeta)\,\mathrm{d}\xi\,\mathrm{d}\zeta. \tag{3.25}
$$

We are free to choose a boundary condition on Σ^- for ϕ^-. The most obvious choice is the free surface condition (2.35). In this case the contribution of (3.25) becomes zero. This results in an auxiliary inner problem with sloshing modes for both the homogeneous Neumann and Dirichlet problems. For a simple object, such as, a rectangular barge we can determine the frequencies of the eigen-modes for both cases explicitly.

3.2.2 *Direct Method*

We consider the Neumann problem with

$$\frac{\partial\phi(\boldsymbol{x})}{\partial n_{\boldsymbol{x}}} = -\frac{\partial\phi_{inc}(\boldsymbol{x})}{\partial n_{\boldsymbol{x}}} \quad \text{for } \boldsymbol{x} \in \mathscr{S}.$$

Similarly as in the acoustic problem, the second relation of (3.22) may be used as an integral equation for ϕ on $\mathscr{S}$:

$$\frac{\phi(\boldsymbol{x})}{2} = -\iint\limits_{\mathscr{S}} \left(\phi(\boldsymbol{\xi})\frac{\partial G(\boldsymbol{x},\boldsymbol{\xi})}{\partial n_{\boldsymbol{\xi}}} + G(\boldsymbol{x},\boldsymbol{\xi})\frac{\partial\phi_{inc}(\boldsymbol{\xi})}{\partial n_{\boldsymbol{\xi}}}\right) \mathrm{d}s_{\boldsymbol{\xi}} \quad \text{for } \boldsymbol{x} \in \mathscr{S}. \tag{3.26}$$

The solution of this equation is not unique for certain irregular frequencies, associated with a resonant inner problem. This may be avoided by means of a different choice of the Green's function. In this case the Green's function may be extended by poles and multi-poles inside the object. There is extensive literature available on this approach in the acoustic case, however, we do not pursue this approach. In the description of the source distribution a different approach is possible.

3.2.3 *Source Distribution*

We have seen for the acoustic problem that we may obtain the solution in terms of a source distribution by adding the two expressions in (3.22) and (3.23). We are free to choose a boundary condition on $\mathscr{S}$ for the inner potential ϕ^-. If we choose $\phi^- = \phi$ and define $\sigma = \frac{\partial\phi}{\partial n_{\xi}} - \frac{\partial\phi^-}{\partial n_{\xi}}$ on $\mathscr{S}$ we obtain

$$\begin{aligned} \phi^-(\boldsymbol{x}) &= \iint\limits_{\mathscr{S}} \sigma(\boldsymbol{\xi})G(\boldsymbol{x},\boldsymbol{\xi})\,\mathrm{d}s_{\boldsymbol{\xi}} + \mathscr{J}^{\Sigma} \quad \text{for } \boldsymbol{x} \in \mathscr{D}^-, \\ \phi(\boldsymbol{x}) &= \iint\limits_{\mathscr{S}} \sigma(\boldsymbol{\xi})G(\boldsymbol{x},\boldsymbol{\xi})\,\mathrm{d}s_{\boldsymbol{\xi}} + \mathscr{J}^{\Sigma} \quad \text{for } \boldsymbol{x} \in \mathscr{D}^+ \cup \mathscr{S}. \end{aligned} \tag{3.27}$$

If we choose the free surface condition for the inner potential we have

$$\phi(\boldsymbol{x}) = \iint\limits_{\mathscr{S}} \sigma(\boldsymbol{\xi})G(\boldsymbol{x},\boldsymbol{\xi})\,\mathrm{d}s_{\boldsymbol{\xi}} \quad \text{for } \boldsymbol{x} \in \mathscr{D}^+ \cup \mathscr{S}. \tag{3.28}$$

For ϕ we have the Neumann condition on $\mathscr{S}$,

$$\frac{\partial\phi(\boldsymbol{x})}{\partial n_{\boldsymbol{x}}} = -\frac{\partial\phi_{inc}(\boldsymbol{x})}{\partial n_{\boldsymbol{x}}}. \tag{3.29}$$

Hence we obtain the following integral equation for the source strength σ:

$$-\frac{\partial \phi_{\mathrm{inc}}(\boldsymbol{x})}{\partial n_{\boldsymbol{x}}} = \frac{1}{2}\sigma(\boldsymbol{x}) + \iint\limits_{\mathscr{S}} \sigma(\boldsymbol{\xi}) \frac{\partial G(\boldsymbol{x}, \boldsymbol{\xi})}{\partial n_{\boldsymbol{x}}} \, \mathrm{d}s_{\boldsymbol{\xi}} \quad \text{for } \boldsymbol{x} \in \mathscr{S}. \tag{3.30}$$

It is easy to show that, for frequencies for which the internal Dirichlet problem has an eigen solution, hence a sloshing mode, this integral equation becomes singular and has no unique solution. Hence the simple source distribution along the hull may not be used at those frequencies.

It is possible to find a unique solution by a different choice of the condition at $y = 0$ for the inner potential ϕ^-. For instance we choose the homogeneous Neumann condition $\frac{\partial \phi^-}{\partial y} = 0$ where the inner problem has a unique solution for all frequencies. We obtain a combined set of equations for the source strength σ on $\mathscr{S}$ and the function ϕ^- at $y = 0$.

$$\begin{aligned}
-\frac{\partial \phi_{inc}(\boldsymbol{x})}{\partial n_{\boldsymbol{x}}} &= \frac{1}{2}\sigma(\boldsymbol{x}) + \iint\limits_{\mathscr{S}} \sigma(\boldsymbol{\xi}) \frac{\partial G(\boldsymbol{x}, \boldsymbol{\xi})}{\partial n_{\boldsymbol{x}}} \, \mathrm{d}s_{\boldsymbol{\xi}} \\
&\quad + \iint\limits_{\Sigma^-} \frac{\omega^2}{g} \phi^-(\xi, 0, \zeta) \frac{\partial G(\boldsymbol{x}, \xi, 0, \zeta)}{\partial n_{\boldsymbol{x}}} \, \mathrm{d}\xi \, \mathrm{d}\zeta \quad \text{for } \boldsymbol{x} \in \mathscr{S}, \\
\phi^-(\boldsymbol{x}) &= \iint\limits_{\mathscr{S}} \sigma(\xi) G(\boldsymbol{x}, \xi) \, \mathrm{d}s_{\xi} \\
&\quad + \iint\limits_{\Sigma^-} \frac{\omega^2}{g} \phi^-(\xi, 0, \zeta) G(\boldsymbol{x}, \xi, 0, \zeta) \, \mathrm{d}\xi \, \mathrm{d}\zeta \quad \text{for } \boldsymbol{x} \in \Sigma^-.
\end{aligned} \tag{3.31}$$

If the exterior problem has a unique solution, it will be obtained by solving this set of equations. Then the solution of this set of equations is unique for all frequencies. It is known that some extremely artificial shapes don't have a unique solution. The hull forms used in ship design and offshore applications generally have unique solutions.

3.2.4 *Motions of a Floating Object, Ship Motions*

If the body moves in one of its six $(i = 1, 2, \ldots, 6)$ degrees of freedom, the normal velocity at the hull of the body is given by

$$\nabla \phi^{(i)} \cdot \mathbf{n} = V^{(i)} \quad \text{for } \boldsymbol{x} \in \mathscr{S}. \tag{3.32}$$

The motion of the object consists of a translation $\boldsymbol{X}$ of the centre of gravity of the object and a rotational motion $\boldsymbol{\Omega}$ relative to the centre of gravity $\boldsymbol{x}_g$ of the body. $V^{(i)}$ corresponds to the translation components $\boldsymbol{X}$ for $(i = 1, 2, 3)$ and to the rotational components $\boldsymbol{\Omega}$ for $(i = 4, 5, 6)$. The combined displacement vector is given by $\boldsymbol{\alpha} = \boldsymbol{X} + \boldsymbol{\Omega} \times (\boldsymbol{x} - \boldsymbol{x}_g) = \tilde{\boldsymbol{\alpha}} \exp(\mathrm{i}\omega t)$ for harmonic motions. In general notation we write

$$\frac{\partial \phi^{(i)}}{\partial n} = V^{(i)} = \mathrm{i}\omega \tilde{\boldsymbol{\alpha}}^{(i)} \cdot \mathbf{n}. \tag{3.33}$$

If we write the potential function $\phi^{(i)}(\boldsymbol{x})$ as a source distribution, with strength $\sigma^{(i)}(\boldsymbol{\xi})$, over S and in the case of an irregular frequency an integral $\mathscr{J}^{\Sigma}$ over the plane $y = 0$ inside the body, we obtain the set of (3.31) with $-\frac{\partial \phi_{inc}(\boldsymbol{x})}{\partial n_{\boldsymbol{x}}}$ replaced by $V^{(i)}$. For the potential of the incident wave we often use the notation $\phi^{(0)}(\boldsymbol{x})$ and for the scattering potential $\phi^{(7)}(\boldsymbol{x})$. The pressure distribution, due to the harmonic potential function (3.19), along the floating fixed body may be calculated by means of the linearised version of the Bernoulli equation (1.13),

$$\frac{p(x, y, z, t)}{\rho} = -\varphi_t(x, y, z, t) = -\mathrm{i}\omega(\phi_0(x, y, z) + \phi_7(x, y, z))\mathrm{e}^{\mathrm{i}\omega t}. \tag{3.34}$$

The linearised harmonic exciting force and moment can now be computed by means of integration of the pressure along the hull. The oscillating hydrodynamic excitation forces and moments are

$$\begin{aligned}
\widetilde{\mathbf{F}}_{exc} &= \mathbf{F}_{exc}\mathrm{e}^{\mathrm{i}\omega t} = -\iint_S p\boldsymbol{n}\,\mathrm{d}S = \rho \mathrm{i}\omega \mathrm{e}^{\mathrm{i}\omega t} \iint_S (\phi_0(x, y, z) + \phi_7(x, y, z))\boldsymbol{n}\,\mathrm{d}S, \\
\widetilde{\mathbf{M}}_{exc} &= \mathbf{M}_{exc}\mathrm{e}^{\mathrm{i}\omega t} = -\iint_S p(\boldsymbol{x} - \boldsymbol{x_g}) \times \boldsymbol{n}\,\mathrm{d}S \\
&= \rho \mathrm{i}\omega \mathrm{e}^{\mathrm{i}\omega t} \iint_S (\phi_0(x, y, z) + \phi_7(x, y, z))(\boldsymbol{x} - \boldsymbol{x_g}) \times \boldsymbol{n}\,\mathrm{d}S.
\end{aligned} \tag{3.35}$$

The harmonic motion of the object gives rise to reaction forces and moments. For $j = 1, \ldots, 6$ we obtain

$$\begin{aligned}
\widetilde{\mathbf{F}}_j &= \mathbf{F}_j\mathrm{e}^{\mathrm{i}\omega t} = -\iint_S p\boldsymbol{n}\,\mathrm{d}S = \rho \mathrm{i}\omega \mathrm{e}^{\mathrm{i}\omega t} \iint_S \phi_j(x, y, z)\boldsymbol{n}\,\mathrm{d}S, \\
\widetilde{\mathbf{M}}_j &= \mathbf{M}_j\mathrm{e}^{\mathrm{i}\omega t} \\
&= -\iint_S p(\boldsymbol{x} - \boldsymbol{x_g}) \times \boldsymbol{n}\,\mathrm{d}S = \rho \mathrm{i}\omega \mathrm{e}^{\mathrm{i}\omega t} \iint_S \phi_j(x, y, z)(\boldsymbol{x} - \boldsymbol{x_g}) \times \boldsymbol{n}\,\mathrm{d}S.
\end{aligned} \tag{3.36}$$

The motion is described by Newton's second law,

$$(-\omega^2 \mathscr{M} + \mathscr{C})\boldsymbol{Y} = \begin{pmatrix} \mathbf{F} \\ \mathbf{M} \end{pmatrix} + \begin{pmatrix} \mathbf{F}_{exc} \\ \mathbf{M}_{exc} \end{pmatrix} \tag{3.37}$$

where $\boldsymbol{Y} = (X_1, X_2, X_3, \Omega_1, \Omega_2, \Omega_3)^T$ is a vector containing the first-order complex translation and rotation amplitudes. $\mathscr{M}$ is the mass matrix containing the mass of the ship and the moments of inertia. The reaction force due to a constant displacement $\boldsymbol{Y}$ is described by $\mathscr{C}\boldsymbol{Y}$, hence the matrix $\mathscr{C}$ may be referred to as the linear *spring* matrix. The forces and moments also depend on the motion, as can be seen in (3.36). We rewrite this part of the force in terms of the added mass and wave damping matrices. If we write

$$\{F_{ij}\} = \begin{pmatrix} \mathbf{F} \\ \mathbf{M} \end{pmatrix},$$

we define the added mass matrix $\mathscr{A}(\omega)$ and wave damping matrix $\mathscr{B}(\omega)$, with entries A_{ij} and B_{ij} respectively, as follows:

$$\begin{pmatrix} \mathbf{F} \\ \mathbf{M} \end{pmatrix} = (\omega^2 \mathscr{A}(\omega) - \mathrm{i}\omega\mathscr{B}(\omega))\boldsymbol{Y}. \tag{3.38}$$

In the frequency domain the equations of motion can now be written as

$$\{-\omega^2(\mathscr{M} + \mathscr{A}(\omega)) + \mathrm{i}\omega\mathscr{B}(\omega) + \mathscr{C}\}\boldsymbol{Y} = \begin{pmatrix} \mathbf{F}_{exc} \\ \mathbf{M}_{exc} \end{pmatrix}. \tag{3.39}$$

The coefficients of this set of equations depend on ω. In the time-domain, the equation for non-harmonic motions may be written as

$$(\mathscr{M} + \overline{\mathscr{A}})\frac{\partial^2 \widetilde{\boldsymbol{Y}}}{\partial t^2} + \mathscr{C}\widetilde{\boldsymbol{Y}} + \int_0^t \mathscr{K}(t-s)\frac{\partial \widetilde{\boldsymbol{Y}}}{\partial s}(s)\,\mathrm{d}s = \begin{pmatrix} \widetilde{\mathbf{F}}_{exc} \\ \widetilde{\mathbf{M}}_{exc} \end{pmatrix}. \tag{3.40}$$

In this equation $\mathscr{K}(t)$ is the step-response matrix, whose entries are the step-response functions K_{ij}. These are oscillating, rapidly-decaying functions which account for the memory part of the equation of motion. The relation between frequency-dependent added mass and damping coefficients and the step-response functions are

$$\begin{aligned} \mathscr{A}(\omega) &= \overline{\mathscr{A}} - \frac{1}{\omega}\int_0^\infty \mathscr{K}(t)\sin(\omega t)\,\mathrm{d}t, \\ \mathscr{B}(\omega) &= \int_0^\infty \mathscr{K}(t)\cos(\omega t)\,\mathrm{d}t. \end{aligned} \tag{3.41}$$

The frequency-dependent damping is the Fourier-cosine transform of the step response function, hence the inverse transform leads to the relation

$$\mathscr{K}(t) = \frac{2}{\pi}\int_0^\infty \mathscr{B}(\omega)\cos(\omega t)\,\mathrm{d}\omega. \tag{3.42}$$

The use of this method to calculate the step-response function is very sensitive to the accuracy of the damping coefficients at high and small values of the frequency. The use of asymptotic expansions is recommended [15].

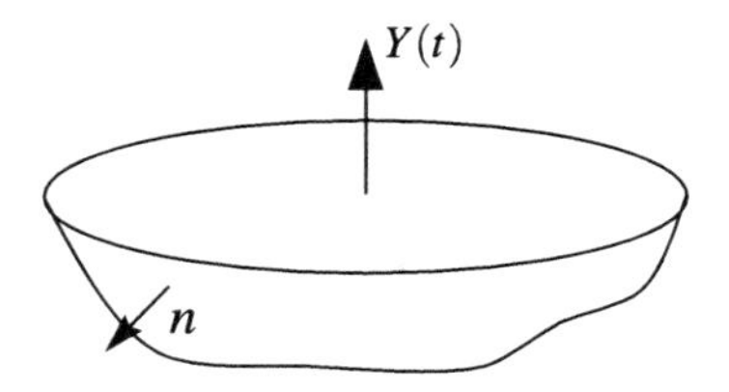

Fig. 3.4 Heave motion

3.2.5 Heave Motion of a Floating Object

We will justify the structure of (3.40) by a direct formulation of an initial value problem. We consider a floating object in infinitely deep still water. At $t = 0$ it is forced to move in one degree of freedom, namely the vertical direction see Fig. 3.4. This restriction is not essential, the derivation given below can easily be extended to six degrees of freedom. The vertical motion is defined as follows:

$$\begin{aligned} \boldsymbol{x}_g &= (0, 0, 0) && \text{for } t \leq 0, \\ \boldsymbol{x}_g &= (0, Y(t), 0) && \text{for } t > 0. \end{aligned} \tag{3.43}$$

We consider the infinitely deep water case, hence the problem to be solved is

$$\begin{aligned} &\varphi_{xx} + \varphi_{yy} + \varphi_{zz} = 0, && \\ &\varphi_{tt} + g\varphi_y = 0, && \text{at } y = 0, \\ &\varphi \to 0, && \text{as } y \to -\infty, \\ &\nabla\varphi \cdot \boldsymbol{n} = \dot{Y}(t) n_2 = f(\boldsymbol{x}, t), && \text{at } S. \end{aligned} \tag{3.44}$$

We assume that the system is at rest for $t \leq 0$ and apply the Laplace transform

$$\Phi(\boldsymbol{x}, s) = \int_0^\infty \mathrm{e}^{-st} \varphi(\boldsymbol{x}, t)\, \mathrm{d}t.$$

The Laplace operator remains unchanged by this transformation. The free surface and boundary condition are transformed in

$$\begin{aligned} &s^2\Phi + g\Phi_y = 0, && \text{at } y = 0, \\ &\nabla\Phi \cdot \boldsymbol{n} = F(\boldsymbol{x}, s), && \text{at } S. \end{aligned} \tag{3.45}$$

To obtain the solution $\Phi(\boldsymbol{x}, s)$ we first consider the source function for this problem. The source function, which obeys the free surface condition in (3.45), follows from (2.77). If we insert $\nu = -\frac{s^2}{g}$ in (2.77) we obtain for a source function with unit strength

$$\Phi_s(\boldsymbol{x}, s) = -\frac{1}{4\pi r} + \frac{1}{4\pi \bar{r}} - \frac{1}{2\pi} \int_0^\infty \frac{gk}{gk + s^2} \mathrm{e}^{k(y+y_0)} J_0(kR)\, \mathrm{d}k. \tag{3.46}$$

We consider a source in the Laplace plane with strength $\overline{\sigma}(s)$. In the time domain this becomes a source with strength $\sigma(t)$, at a fixed position, and it is obtained by the inverse Laplace transform of $\overline{\sigma}(s)\Phi_s(\boldsymbol{x}, s)$. We obtain

$$\varphi_s(\boldsymbol{x}, t) = -\frac{\sigma(t)}{4\pi r} + \frac{\sigma(t)}{4\pi \bar{r}} - \frac{1}{2\pi}\int_0^\infty \int_0^t \sigma(\tau)\sqrt{gk}\sin(\sqrt{gk}(t-\tau))\mathrm{e}^{k(y+y_0)} J_0(kR)\,\mathrm{d}\tau\,\mathrm{d}k. \quad (3.47)$$

The first part of (3.47) is in accordance with the result (13.49) in the book of Wehausen [19]. This source function may be used in the application of Green's theorem to obtain an integral equation for the time dependent potential function $\varphi(\boldsymbol{x}, t)$. However for our purpose it is more convenient to define a Green's function $G(\boldsymbol{x}, \boldsymbol{\xi}, s)$, where $\boldsymbol{x} = \boldsymbol{\xi}$ is the position of the singularity, in the Laplace domain, which fulfils the homogeneous boundary condition at the surface S of the floating object:

$$\nabla G \cdot \boldsymbol{n} = 0, \quad \text{at } S. \quad (3.48)$$

Application of Green's theorem gives

$$\Phi(\boldsymbol{x}, s) = \iint_S G(\boldsymbol{x}, \boldsymbol{\xi}, s) F(\boldsymbol{\xi}, s)\,\mathrm{d}S_{\xi} = V(s)\iint_S G(\boldsymbol{x}, \boldsymbol{\xi}, s) n_2(\boldsymbol{\xi})\,\mathrm{d}S_{\xi}, \quad (3.49)$$

where $V(s)$ is the Laplace transform of the vertical motion $\dot{Y}(t)$. One must realise that the Green's function G can not be determined explicitly. However the reason that it is introduced is that we wish to verify the structure of the equation of motion (3.40). To do so, we define G_a as

$$G_a = \frac{1}{4\pi}\left\{-\frac{1}{|\boldsymbol{x} - \boldsymbol{\xi}|} + \frac{1}{|\boldsymbol{x} - \bar{\boldsymbol{\xi}}|}\right\} + g_a(\boldsymbol{x}, \boldsymbol{\xi}, s) \quad (3.50)$$

where $g_a = 0$ at $y = 0$ and $\frac{\partial G_a}{\partial n} = 0$ at the surface S. This gives a unique decomposition of G, i.e.,

$$G(\boldsymbol{x}, \boldsymbol{\xi}, \boldsymbol{s}) = G_a(\boldsymbol{x}, \boldsymbol{\xi}) + K_a(\boldsymbol{x}, \boldsymbol{\xi}, \boldsymbol{s}). \quad (3.51)$$

We are now able to rewrite (3.49) in the form

$$\Phi(\boldsymbol{x}, s) = V(s)\iint_S G_a(\boldsymbol{x}, \boldsymbol{\xi}) n_2(\boldsymbol{\xi})\,\mathrm{d}S_{\xi} + V(s)\iint_S K_a(\boldsymbol{x}, \boldsymbol{\xi}, s) n_2(\boldsymbol{\xi})\,\mathrm{d}S_{\xi}. \quad (3.52)$$

The inverse Laplace transform of (3.52) gives

$$\varphi(\boldsymbol{x}, t) = \dot{Y}(t)\psi(\boldsymbol{x}) + \int_0^t \chi(\boldsymbol{x}, t-\tau)\dot{Y}(\tau)\,\mathrm{d}\tau \quad (3.53)$$

where

$$\psi(\boldsymbol{x}) = \iint\limits_{S} G_a(\boldsymbol{x}, \boldsymbol{\xi}) n_2(\boldsymbol{\xi})\, \mathrm{d}S_{\xi} \tag{3.54}$$

and

$$\chi(t) = \iint\limits_{S} k_a(\boldsymbol{x}, \boldsymbol{\xi}, t) n_2(\boldsymbol{\xi})\, \mathrm{d}S_{\xi} \tag{3.55}$$

where $k_a(\boldsymbol{x}, \boldsymbol{\xi}, t)$ is the inverse Laplace transform of $K_a(\boldsymbol{x}, \boldsymbol{\xi}, s)$. We are now able to compute the pressure along the hull S.

$$p(\boldsymbol{x}, t) = -\frac{1}{\rho}\varphi_t(\boldsymbol{x}, t) = -\frac{1}{\rho}\left\{\ddot{Y}(t)\psi(\boldsymbol{x}) + \int_0^t \dot{\chi}(\boldsymbol{x}, t-\tau)\dot{Y}(\tau)\, \mathrm{d}\tau\right\}. \tag{3.56}$$

Next we may compute the forces and moments acting on the hull. We restrict ourselves to the heave force, while the other components can be determined in the same manner,

$$F_2(t) = -\frac{1}{\rho}\iint\limits_{S} \varphi_t(\boldsymbol{x}) n_2(\boldsymbol{x})\, \mathrm{d}S = \frac{1}{\rho}\left\{\ddot{Y}(t)\iint\limits_{S} \psi(\boldsymbol{x}) n_2(\boldsymbol{x})\, \mathrm{d}S \right.$$
$$\left. + \iint\limits_{S}\int_0^t \dot{\chi}(\boldsymbol{x}, t-\tau)\dot{Y}(\tau) n_2(\boldsymbol{x})\, \mathrm{d}\tau\, \mathrm{d}S\right\}. \tag{3.57}$$

The Newton equation for the heave motion now becomes

$$M\frac{\mathrm{d}^2 Y(t)}{\mathrm{d}t^2} + c_{22}Y(t) = F_2(t) + F_{ext}(t), \tag{3.58}$$

where $F_{ext}(t)$ is an external force acting on the object. If we compare result (3.57) with (3.40) for the heave motion

$$(M + \bar{a}_{22})\frac{\mathrm{d}^2 Y(t)}{\mathrm{d}t^2} + c_{22}Y(t) + \int_0^t K_{22}(t-\tau)\frac{\mathrm{d}Y(\tau)}{\mathrm{d}\tau}\, \mathrm{d}\tau = F_{ext}(t), \tag{3.59}$$

we see that

$$\begin{aligned} \bar{a}_{22} &= -\frac{1}{\rho}\iint\limits_{S} \psi(\boldsymbol{x}) n_2(\boldsymbol{x})\, \mathrm{d}S, \\ K_{22}(t) &= -\frac{1}{\rho}\iint\limits_{S} \dot{\chi}(\boldsymbol{x}, t) n_2(\boldsymbol{x})\, \mathrm{d}S. \end{aligned} \tag{3.60}$$

In a similar way we may obtain results for the motion in six degrees of freedom to justify the structure of (3.40) completely, see [14]. The equations for the ship motions for a ship sailing in waves can be obtained as long as the steady part of the potential can be described by Ux, which means for thin or slender ships.

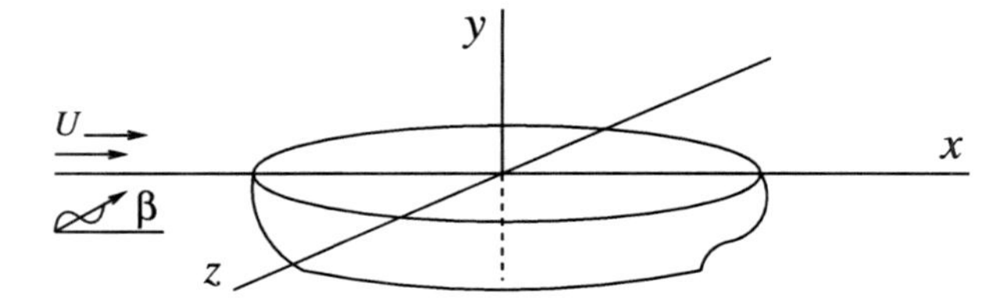

Fig. 3.5 Coordinate system

3.3 Slow Speed Approximation

We first derive the equations for the potential function $\varphi(\mathbf{x}, t)$, such that the fluid velocity $\mathbf{u}(\mathbf{x}, t)$ is defined as $\mathbf{u}(\mathbf{x}, t) = \nabla\varphi(\mathbf{x}, t)$. The total potential function will be split up in a steady and a non-steady part in the following way:

$$\varphi(\mathbf{x}, t) = Ux + \overline{\phi}(\mathbf{x}; U) + \tilde{\phi}(\mathbf{x}, t; U).$$

In this formulation U is the incoming unperturbed velocity field, obtained by considering a coordinate system fixed to the ship moving under a drift angle β see Fig. 3.5. In our approach this angle need not be small. The time dependent part of the potential consists of an incoming, diffracted and radiated (for the six modes of motion) wave

$$\tilde{\phi}(\mathbf{x}, t; U) = \tilde{\phi}^{inc}(\mathbf{x}, t; U) + \tilde{\phi}^{(7)}(\mathbf{x}, t; U) + \sum_{j=1}^{6} \tilde{\phi}^{(j)}(\mathbf{x}, t; U)$$

at frequency $\omega = \omega_0 + k_0 U \cos\beta$, where ω_0 and $k_0 = \omega_0^2/g$ are the frequency and wave number in the earth-fixed coordinate system, while ω is the frequency in the coordinate system fixed to the ship. The waves are incoming under an angle β, with respect to the current. To compute the linear wave forces all these components will be taken into account.

We consider the deep water case, hence $h = -\infty$ and the equations for the total potential φ can be written as:

$$\Delta\varphi = 0 \text{ in the fluid domain } D_e.$$

At the free surface we have the dynamic and kinematic boundary condition

$$\left.\begin{aligned} g\eta + \varphi_t + \frac{1}{2}\nabla\varphi \cdot \nabla\varphi &= \mathscr{C}_{st} \\ \varphi_y - \varphi_x \eta_x - \varphi_z \eta_z - \eta_t &= 0 \end{aligned}\right\} \quad \text{at } y = \eta. \tag{3.61}$$

At the body surface we have

$$\frac{\partial\varphi}{\partial n} = \mathbf{V} \cdot \mathbf{n}$$

where $\mathbf{V}$ is the body velocity relative to the average body-fixed coordinate system.

We assume that the waves are high compared to the Kelvin stationary wave pattern, but that they are both small in nature, hence the free surface boundary condition can be expanded at $y = 0$. We first eliminate η, which leads to the following boundary condition:

$$\frac{\partial^2}{\partial t^2}\varphi + g\frac{\partial}{\partial y}\varphi + \frac{\partial}{\partial t}(\nabla\varphi \cdot \nabla\varphi) + \frac{1}{2}\nabla\varphi \cdot \nabla[\nabla\varphi \cdot \nabla\varphi] = 0 \quad \text{at } y = \eta. \tag{3.62}$$

To compute the first-order wave potential the free surface has to be linearised first. We assume $\tilde{\phi}(\mathbf{x}, t; U) = \phi(\mathbf{x}; U)\exp(\mathrm{i}\omega t)$, then for each degree of freedom, $i = 1, \ldots, 6$, the free surface condition at $y = 0$ can be written as

$$-\omega^2\phi + 2\mathrm{i}\omega U\phi_x + U^2\phi_{xx} + g\phi_y = \mathrm{i}\omega\mathscr{D}(U; \overline{\phi})\{\phi\} \quad \text{at } y = 0. \tag{3.63}$$

while for the diffracted potential $\phi^{(7)}$ the last term has to be replaced by $\mathscr{D}(U; \overline{\phi})\{\phi^{inc} + \phi^{(7)}\}$ and where $\mathscr{D}(U; \overline{\phi})$ is the following linear differential operator acting on ϕ. The quadratic terms in ϕ are neglected.

$$\begin{aligned}
\mathrm{i}\omega\mathscr{D}(U; \overline{\phi})\{\phi\} = {} & \mathrm{i}\omega\{(\overline{\phi}_{xx} + \overline{\phi}_{zz})\phi + 2\nabla\overline{\phi} \cdot \nabla\phi\} \\
& + (2U\overline{\phi}_x + \overline{\phi^2}_x)\phi_{xx} + 2(U + \overline{\phi}_x)\overline{\phi}_z\phi_{xz} + \overline{\phi^2}_z\phi_{zz} \\
& + (3U\overline{\phi}_{xx} + \overline{\phi}_x\overline{\phi}_{xx} + \overline{\phi}_z\overline{\phi}_{xz})\phi_x + (2U\overline{\phi}_{xz} + \overline{\phi}_x\overline{\phi}_{xz} \\
& + \overline{\phi}_z\overline{\phi}_{zz})\phi_z.
\end{aligned}$$

The linear problems for $\phi^{(j)}$ with $j = 1, \ldots, 7$ are solved by means of a source distribution along the ship hull, its water line and the free surface $y = 0$. We write for each potential function:

$$\begin{aligned}
4\pi\phi(\mathbf{x}) = {} & \iint_S \sigma(\boldsymbol{\xi})G(\mathbf{x}, \boldsymbol{\xi})\,\mathrm{d}S_\xi - \frac{U^2}{g}\int_{WL} \alpha_n\sigma(\boldsymbol{\xi})G(\mathbf{x}, \boldsymbol{\xi})\,\mathrm{d}s_\xi \\
& + \frac{\mathrm{i}\omega}{g}\iint_{FS} G(\mathbf{x}, \boldsymbol{\xi})\mathscr{D}\{\phi\}\,\mathrm{d}S_\xi \quad \text{for } \mathbf{x} \in D_e.
\end{aligned} \tag{3.64}$$

The distribution along the water line follows after the use of the divergence theorem in the integral at the free surface $y = 0$. The function $G(\mathbf{x}, \boldsymbol{\xi})$ is the Green's function that obeys the free surface condition (3.63) with $\mathscr{D}$ equals zero and $\alpha_n = e_x \cdot \mathbf{n}$, where e_x equals the unit vector in the x-direction. In general the boundary conditions on the ship are given in the form

$$\nabla\phi^{(j)} \cdot \mathbf{n} = V^{(j)}(\mathbf{x}) \quad \text{for } \mathbf{x} \in S \text{ and } j = 1, \ldots, 7,$$

where $V^{(j)}$ is the normal velocity due to the motion in the jth mode with $j = 1, \ldots, 6$ and $V^{(7)}$ equals the normal velocity of the incident oscillating field. So $V^{(j)}$, with $j = 1, \ldots, 3$, correspond to the translation components $\mathbf{X} = \tilde{\mathbf{X}}\exp(\mathrm{i}\omega t)$ and, with $j = 4, \ldots, 6$, to components of the rotational motion $\boldsymbol{\Omega} = \tilde{\boldsymbol{\Omega}}\exp(\mathrm{i}\omega t)$ relative to the centre of gravity $\mathbf{x}_g$ of the body. The combined displacement vector is given by $\boldsymbol{\alpha} = \mathbf{X} + \boldsymbol{\Omega} \times (\mathbf{x} - \mathbf{x}_g) = \tilde{\boldsymbol{\alpha}}\exp(\mathrm{i}\omega t)$. In general notation we write

$$\frac{\partial\phi}{\partial n} = \mathrm{i}\omega\tilde{\boldsymbol{\alpha}} \cdot \mathbf{n} + [(\nabla(Ux + \overline{\phi}) \cdot \nabla)\tilde{\boldsymbol{\alpha}} - (\tilde{\boldsymbol{\alpha}} \cdot \nabla)\nabla(Ux + \overline{\phi})] \cdot \mathbf{n}.$$

This leads to an equation for the source strength, where we omitted the index j again:

$$-2\pi\sigma(\mathbf{x}) - \iint_S \sigma(\boldsymbol{\xi})\frac{\partial}{\partial \mathbf{n}_x}G(\mathbf{x},\boldsymbol{\xi})\,\mathrm{d}S_\xi + \frac{U^2}{g}\int_{WL}\alpha_n\sigma(\boldsymbol{\xi})\frac{\partial}{\partial \mathbf{n}_x}G(\mathbf{x},\boldsymbol{\xi})\,\mathrm{d}s_\xi$$
$$+\frac{\mathrm{i}\omega}{g}\iint_{FS}\frac{\partial}{\partial \mathbf{n}_x}G(\mathbf{x},\boldsymbol{\xi})\mathscr{D}\{\phi\}\,\mathrm{d}S_\xi = 4\pi V(\mathbf{x}) \quad \text{for } \mathbf{x}\in S. \tag{3.65}$$

If the forward velocity U is small we can simplify the $\mathscr{D}$ and solve this equation iteratively. However an accurate numerical evaluation of the complete Green's function is rather elaborate. Therefore we could also make use of the fact that U is small, keeping in mind that there are two dimensionless parameters that play a role, namely $\tau = \frac{\omega U}{g} \ll 1$ and $\nu = \frac{gL}{U^2} \gg 1$. The source potentials and the strengths can be evaluated as perturbation series with respect to τ,

$$\sigma(\boldsymbol{\xi}) = \sigma_0(\boldsymbol{\xi}) + \tau\sigma_1(\boldsymbol{\xi}) + \hat{\sigma}(\boldsymbol{\xi}; U), \tag{3.66}$$

$$\phi(\mathbf{x}) = \phi_0(\mathbf{x}) + \tau\phi_1(\mathbf{x}) + \hat{\phi}(\mathbf{x}; U), \tag{3.67}$$

where $\hat{\sigma}$ and $\hat{\phi}$ are $O(\tau^2)$ as $\tau \to 0$. The expansion of G is less trivial. We write

$$G(\mathbf{x},\boldsymbol{\xi};U) = -\frac{1}{\mathbf{r}} + \frac{1}{\mathbf{r}'} - \{\psi_0(\mathbf{x},\boldsymbol{\xi}) + \tau\psi_1(\mathbf{x},\boldsymbol{\xi}) + \cdots$$
$$+\tilde{\psi}_0(\mathbf{x},\boldsymbol{\xi}) + \nu^{-1}\tilde{\psi}_1(\mathbf{x},\boldsymbol{\xi}) + \cdots\}. \tag{3.68}$$

The first term between brackets corresponds to the Green's function at zero forward speed, for which there exist several fast computer codes. The second term is the modification due to small values of the forward velocity. Computations can be carried out by means of a modification of the existing fast code. The disadvantage of this approach is the fact that the asymptotic expansion of the Green's function is only valid close to the source. At finite distance the nonuniform asymptotic behaviour has to be corrected. The third term between brackets is the one that describes the Kelvin effect on the wave Green's function. Although this term is linear in ν and therefore tends to infinity as U goes to zero. Because it becomes highly oscillatory it can be shown that its contribution to the potential function becomes zero. In practice the first two terms are computed in the expansions of the potentials and the source strengths for the excitation and the six modes of the motion. This approach can be applied to the situation of deep water. For finite water depth the evaluation of the Green's function for finite velocities leads to terms that are not as easy to compute as in the deep water case, where all the expressions needed can be expressed in derivatives of the zero speed Green's function. After all this approach of splitting of the Green's function is not very practical for real ship forms. At this moment there are some codes available to compute the total Green's function directly in an efficient way.

Chapter 4
Second-Order Theory

For some applications it is worthwhile to consider second-order wave effects. Especially the influence of low frequency second-order forces on systems moored in waves may cause resonant behaviour, resulting in large oscillating motions at the resonant frequency of the moored system. To describe this phenomenon we first derive the second-order waves in the case that we have a simple spectrum of waves consisting of two monochromatic plane waves with frequencies close to each other. Next a derivation of the second-order low frequency drift forces by means of a local expansion and far-field expansions is given. As a classroom example we consider the forces on a vertical wall caused by the reflection of such a wave system. This serves as an introduction to formulation of the low frequency motion of a moored system.

4.1 Second-Order Wave Theory

The general non-linear free surface conditions are given by (1.11) and (1.14). For two-dimensional plane waves they become

$$\left.\begin{aligned} &\varphi_y = \eta_t + \varphi_x \eta_x, \\ &\varphi_t + \frac{1}{2}(\varphi_x^2 + \varphi_y^2) + g\eta = \text{constant} \end{aligned}\right\} \quad \text{at } y = \eta(x,t). \tag{4.1}$$

We consider the situation where $\eta(x,t)$ and $\varphi(x,y,t)$ are small of $O(\varepsilon)$, where ε is a measure for the wave height. Hence the potential function $\varphi(x,\eta(x,t),t)$ and its derivatives may be expanded in a Taylor series at $y = 0$. To derive a second-order formulation for the free surface condition we need some terms up to second-order in the small parameter η. However the x-derivative is needed in first-order only. We expand φ in the following way:

A.J. Hermans, *Water Waves and Ship Hydrodynamics*,
DOI 10.1007/978-94-007-0096-3_4,

$$\begin{aligned}
\varphi(x,\eta(x,t),t) &= \varphi(x,0,t)+\eta(x,t)\varphi_y(x,0,t)+\mathrm{O}(\varepsilon^2),\\
\varphi_x(x,\eta(x,t),t) &= \varphi_x(x,0,t)+\mathrm{O}(\varepsilon),\\
\varphi_y(x,\eta(x,t),t) &= \varphi_y(x,0,t)+\eta(x,t)\varphi_{yy}(x,0,t)+\mathrm{O}(\varepsilon^2),\\
\varphi_t(x,\eta(x,t),t) &= \varphi_t(x,0,t)+\eta(x,t)\varphi_{yt}(x,0,t)+\mathrm{O}(\varepsilon^2).
\end{aligned} \tag{4.2}$$

Hence we have

$$\begin{aligned}
-g\eta &= \left[\varphi_t+\eta\varphi_{yt}+\frac{1}{2}(\varphi_x^2+\varphi_y^2)\right]_{y=0}+\mathrm{O}(\varepsilon^3)\\
&= \left[\varphi_t-\frac{\varphi_t\varphi_{yt}}{g}+\frac{1}{2}(\varphi_x^2+\varphi_y^2)\right]_{y=0}+\mathrm{O}(\varepsilon^3).
\end{aligned} \tag{4.3}$$

To eliminate η in (4.1) we need its derivatives

$$\begin{aligned}
-g\eta_t &= \left[\varphi_{tt}-\frac{\varphi_{tt}\varphi_{yt}}{g}-\frac{\varphi_t\varphi_{ytt}}{g}+\varphi_x\varphi_{xt}+\varphi_y\varphi_{yt}\right]_{y=0}+\mathrm{O}(\varepsilon^3),\\
-g\eta_x &= \varphi_{xt}|_{y=0}+\mathrm{O}(\varepsilon^2).
\end{aligned} \tag{4.4}$$

The free surface condition can now be written at $y=0$ as

$$g\varphi_y+\varphi_{tt}=\varphi_{yy}\varphi_t-2\varphi_x\varphi_{xt}-\varphi_y\varphi_{yt}+\frac{\varphi_{tt}\varphi_{yt}}{g}+\frac{\varphi_t\varphi_{ytt}}{g}+\mathrm{O}(\varepsilon^3). \tag{4.5}$$

The potential function will be written as an asymptotic expansion with respect to the small parameter ε,

$$\varphi(x,y,t)=\varepsilon\varphi^{(1)}(x,y,t)+\ \varepsilon^2\varphi^{(2)}(x,y,t)+\mathrm{O}(\varepsilon^3). \tag{4.6}$$

We insert this expansion in expression (4.6) and equate equal powers in ε equal to zero. We then obtain conditions for $\varphi^{(i)}$ at $y=0$,

$$\begin{aligned}
g\varphi_y^{(1)}+\varphi_{tt}^{(1)} &= 0,\\
g\varphi_y^{(2)}+\varphi_{tt}^{(2)} &= \varphi_{yy}^{(1)}\varphi_t^{(1)}-2\varphi_x^{(1)}\varphi_{xt}^{(1)}-\varphi_y^{(1)}\varphi_{yt}^{(1)}+\frac{\varphi_{tt}^{(1)}\varphi_{yt}^{(1)}}{g}+\frac{\varphi_t^{(1)}\varphi_{ytt}^{(1)}}{g}\\
&= -2\big(\varphi_x^{(1)}\varphi_{xt}^{(1)}+\varphi_y^{(1)}\varphi_{yt}^{(1)}\big)+\frac{1}{g}\varphi_t^{(1)}\frac{\partial}{\partial y}\big(g\varphi_y^{(1)}+\varphi_{tt}^{(1)}\big),
\end{aligned} \tag{4.7}$$

where we used the expression for $\varphi^{(1)}$ to obtain the expression on the last line.

We consider an incident wave field consisting of two monochromatic plane waves with frequencies, ω and $\omega+\delta\omega$, close to each other. The first-order amplitude may be written as

$$\eta^{(1)}(x,t)=\Re\big\{a_1\mathrm{e}^{\mathrm{i}(\omega t-kx)}\big\}+\Re\big\{a_2\mathrm{e}^{\mathrm{i}((\omega+\delta\omega)t-(k+\delta k)x+\vartheta)}\big\}, \tag{4.8}$$

where we choose a_1 and a_2 to be real constants. For finite water depth the first-order potential function may be written as

$$\varphi^{(1)}(x,y,t) = \Re\{\psi_1(x,y,t) + \psi_2(x,y,t)\} = \Re\{\Psi\} = \frac{\Psi + \overline{\Psi}}{2}, \tag{4.9}$$

with

$$\begin{aligned} \psi_1(x,y,t) &= \alpha_1 \frac{\cosh(k_1(y+h))}{\cosh(k_1 h)} \mathrm{e}^{\mathrm{i}(\omega_1 t - k_1 x)} \quad \text{and} \\ \psi_2(x,y,t) &= \alpha_2 \frac{\cosh(k_2(y+h))}{\cosh(k_2 h)} \mathrm{e}^{\mathrm{i}(\omega_2 t - k_2 x + \vartheta)} \end{aligned} \tag{4.10}$$

where $\alpha_1 = -\frac{\mathrm{i}g}{\omega_1} a_1$ and $\alpha_2 = -\frac{\mathrm{i}g}{\omega_2} a_2$ are imaginary constants. The free surface condition for $\varphi^{(2)}$ may be written as

$$\begin{aligned} g\varphi_y^{(2)} + \varphi_{tt}^{(2)} &= -\frac{1}{2}\{(\Psi_x + \overline{\Psi}_x)(\Psi_{xt} + \overline{\Psi}_{xt}) + (\Psi_y + \overline{\Psi}_y)(\Psi_{yt} + \overline{\Psi}_{yt})\} \\ &\quad + \frac{1}{4g}(\Psi_t + \overline{\Psi}_t)\frac{\partial}{\partial y}\big(\overline{(g\Psi_y + \Psi_{tt})} + (g\Psi_y + \Psi_{tt})\big). \end{aligned} \tag{4.11}$$

If we insert (4.9) in expression (4.10) we notice that several exponential combinations play a role. We see among others if we combine ψ_1 with itself that we obtain a contribution with the double frequency, $2\omega t$. Here we are specially interested in contributions with the difference frequency $\delta\omega = \omega_1 - \omega_2$. These terms are generated by the products of Ψ_1 and $\overline{\Psi}_2$ and of Ψ_2 and $\overline{\Psi}_1$. We restrict ourself to $\delta\omega$ contributions and consider first the terms

$$\begin{aligned} &\Psi_{1x}\overline{\Psi}_{2xt} + \Psi_{2x}\overline{\Psi}_{1xt} + \overline{\Psi}_{1x}\Psi_{2xt} + \overline{\Psi}_{2x}\Psi_{1xt} \\ &\quad = 2k_1k_2\delta\omega\alpha_1\alpha_2\sin(\delta\omega t - \delta k x + \vartheta), \end{aligned} \tag{4.12}$$

with $\delta k = k_1 - k_2$. We also find

$$\begin{aligned} &\Psi_{1y}\overline{\Psi}_{2yt} + \Psi_{2y}\overline{\Psi}_{1yt} + \overline{\Psi}_{1y}\Psi_{2yt} + \overline{\Psi}_{2y}\Psi_{1yt} \\ &\quad = 2k_1k_2\delta\omega\alpha_1\alpha_2\tanh(k_1h)\tanh(k_2h)\sin(\delta\omega t - \delta k x + \vartheta). \end{aligned} \tag{4.13}$$

The slowly varying part in the last term becomes, for small values of the frequency difference $\delta\omega$,

$$\begin{aligned} &\frac{1}{4g}(\Psi_t + \overline{\Psi}_t)\frac{\partial}{\partial y}\big(\overline{(g\Psi_y + \Psi_{tt})} + (g\Psi_y + \Psi_{tt})\big) \\ &\approx -\frac{1}{2}\omega^2\delta\omega\alpha_1\alpha_2\frac{\mathrm{d}}{\mathrm{d}\omega}\left(\frac{\omega^3}{g^2\sinh^2(kh)}\right)\sin(\delta\omega t - \delta k x + \vartheta) + \mathrm{O}\big((\delta\omega)^2\big). \end{aligned} \tag{4.14}$$

Hence the free surface condition, at $y=0$, for the *bound* second-order wave with small frequency $\delta\omega$ becomes

$$g\varphi_y^{(2)} + \varphi_{tt}^{(2)} = \overline{\mathscr{C}}\delta\omega\alpha_1\alpha_2\sin(\delta\omega t - \delta kx + \vartheta), \tag{4.15}$$

with

$$\overline{\mathscr{C}} = -k^2\frac{\cosh(2kh)}{\cosh^2(kh)} - \frac{1}{2}\omega^2\frac{\mathrm{d}}{\mathrm{d}\omega}\left(\frac{\omega^3}{g^2\sinh^2(kh)}\right). \tag{4.16}$$

The solution for the low frequency wave potential becomes

$$\varphi^{(2)}(x,y,t) = \frac{\overline{\mathscr{C}}\delta\omega\alpha_1\alpha_2\cosh(\delta k(y+h))}{g\delta k\sinh(\delta kh) - (\delta\omega)^2\cosh(\delta kh)}\sin(\delta\omega t - \delta kx + \vartheta). \tag{4.17}$$

We rewrite this solution as

$$\varphi^{(2)}(x,y,t) = \Re\left\{\mathscr{C}\alpha_1\alpha_2\cosh(\delta k(y+h))\mathrm{e}^{\mathrm{i}(\delta\omega t - \delta kx + \vartheta)}\right\}, \tag{4.18}$$

where

$$\mathscr{C} = \frac{-\mathrm{i}\overline{\mathscr{C}}\delta\omega}{g\delta k\sinh(\delta kh) - (\delta\omega)^2\cosh(\delta kh)}. \tag{4.19}$$

This represents a slowly varying second-order wave with wave velocity equal to the group velocity $c_g = \frac{\mathrm{d}\omega}{\mathrm{d}k}$. One must realise that this wave does not obey the linear first-order free surface condition. Hence when it is disturbed by a fixed object it generates linear first-order waves at the frequency $\delta\omega$, while the refracted first-order waves generate a *bound* second-order wave as well. So we have to distinguish at least three different types of waves in this case.

4.2 Wave-Drift Forces and Moments

The constant wave drift force acting on a three-dimensional object in waves, while a small current is present, can be computed in two different ways, while for the low frequency wave drift force only one method remains. In both cases we are able to perform a pressure integration along the body, while for the constant term the use of the law of conservation of momentum leads to an expression consisting of the far-field wave pattern only. If we take into account the effect of the current we may write the total potential as follows:

$$\begin{aligned}\varphi(\mathbf{x},t) &= Ux + \overline{\phi}(\mathbf{x};U) + \phi(\mathbf{x};U)\mathrm{e}^{-\mathrm{i}\omega t}\\ &= Ux + \overline{\phi}(\mathbf{x};U) + \left\{\phi^{inc}(\mathbf{x};U) + \phi^{(7)}(\mathbf{x};U) + \sum_{j=1}^{6}\phi^{(j)}(\mathbf{x};U)\overline{\alpha}_j\right\}\mathrm{e}^{-\mathrm{i}\omega t}.\end{aligned} \tag{4.20}$$

By means of this expression we describe the situation of diffraction of a monochromatic incident. The steady disturbance $\overline{\phi}(\mathbf{x}; U)$ is assumed to be small compared with Ux. If this is not the case we also have to take care of the dependency of the time dependent potential function on $\overline{\phi}(\mathbf{x}; U)$. If we want to compute the harmonic potential by means of a superposition of source functions as described before we must ignore the effect of $\overline{\phi}(\mathbf{x}; U)$. If we solve the unsteady potential in the time-domain it can be taken into account. In the formulation for the second-order constant or low frequency forces and moments we can take the effect of $\overline{\phi}(\mathbf{x}; U)$ into account. The potential $\phi^{(7)}(\mathbf{x}; U)$ is the wave scattered by the incident wave field. The potential function $\phi^{(j)}(\mathbf{x}; U)$ is the reaction potential due to the motion, $\overline{\alpha}_j$, of the floating object in one of the six degrees of freedom.

4.2.1 *Constant and Low Frequency Drift Forces by Means of Local Expansions*

First we treat the second-order slowly varying drift forces. The forces acting on the hull can be obtained by integrating the pressure along the exact wetted hull surface $\tilde{S}$. The pressure on the surface is given by Bernoulli's equation

$$p(\mathbf{x}, t) = -\rho\left(\frac{\partial \varphi}{\partial t} + \frac{1}{2}\nabla\varphi \cdot \nabla\varphi + gy - \frac{1}{2}U^2\right) \quad \text{on } \tilde{S}.$$

The force acting on the hull is obtained by integration of the pressure over the exact surface

$$\mathbf{F} = \int_{\tilde{S}} p\mathbf{n}\,\mathrm{d}S. \tag{4.21}$$

This looks like an easy operation, but it is not because the linearised potentials are computed at the mean hull surface and the integration goes unto the free water surface, while liberalisation with respect to $y = 0$ has taken place. Hence the expression for the force has to be expanded to integrals over the mean hull surface S and the unperturbed water-line. We assume that the total displacement of a point $\mathbf{x}$ of the surface is given in linearised form as

$$\boldsymbol{\alpha} = \mathbf{X} + \boldsymbol{\Omega} \times (\mathbf{x} - \mathbf{x}_g) \tag{4.22}$$

with $\mathbf{X}$ the translational and $\boldsymbol{\Omega}$ the rotational motion of the body relative to its centre of gravity $\mathbf{x}_g$. At this point several assumptions are made, some of them being more or less questionable. The pressure is expanded in a Taylor series around the average surface S, which can be done because the pressure is a differentiable function of $\mathbf{x}$.

$$p_{\tilde{S}} = p_S + \boldsymbol{\alpha} \cdot \nabla p_S + \frac{1}{2}(\boldsymbol{\alpha} \cdot \nabla)^2 p_S + \mathcal{O}(|\boldsymbol{\alpha}|^3). \tag{4.23}$$

A step that poses some geometrical limitations, like a vertical ship hull at the free surface, on the applicability of this approach is the evaluation of the integral

$$\int_{\tilde{S}} f(\mathbf{x})\,\mathrm{d}S \approx \int_{S} f(\mathbf{x})\,\mathrm{d}S + \int_{wl}\int_{\alpha_2}^{\eta} f(\mathbf{x})\,\mathrm{d}y\,\mathrm{d}l. \tag{4.24}$$

We assume the height of the waves and the motions of the hull to be small O(ε), hence we expand all quantities with respect to ε as follows:

$$\begin{aligned}
\varphi(\mathbf{x},t) &= \overline{\varphi}(\mathbf{x}) + \varepsilon\psi^{(1)}(\mathbf{x},t) + \varepsilon^2\psi^{(2)}(\mathbf{x},t) + \mathscr{O}(\varepsilon^3),\\
\eta(x,z,t) &= \overline{\eta}(x,z) + \varepsilon\eta^{(1)}(x,z,t) + \varepsilon^2\eta^{(2)}(x,z,t) + \mathscr{O}(\varepsilon^3),\\
p_S &= \overline{p}(\mathbf{x}) + \varepsilon p^{(1)}(\mathbf{x},t) + \varepsilon^2 p^{(2)}(\mathbf{x},t) + \mathscr{O}(\varepsilon^3),\\
\mathbf{X} &= \varepsilon\mathbf{X}^{(1)} + \varepsilon^2\mathbf{X}^{(2)} + \mathscr{O}(\varepsilon^3),\\
\mathbf{\Omega} &= \varepsilon\mathbf{\Omega}^{(1)} + \varepsilon^2\mathbf{\Omega}^{(2)} + \mathscr{O}(\varepsilon^3),\\
\mathbf{x} - \mathbf{x}_g &= \overline{\mathbf{x}} - \mathbf{x}_g + \varepsilon\mathbf{\Omega}^{(1)} \times (\overline{\mathbf{x}} - \mathbf{x}_g) + \varepsilon^2\mathbf{\Omega}^{(2)} \times (\overline{\mathbf{x}} - \mathbf{x}_g) + \mathscr{O}(\varepsilon^3),\\
\mathbf{n} &= \overline{\mathbf{n}} + \varepsilon\mathbf{\Omega}^{(1)} \times \overline{\mathbf{n}} + \varepsilon^2\mathbf{\Omega}^{(2)} \times \overline{\mathbf{n}} + \mathscr{O}(\varepsilon^3),
\end{aligned} \tag{4.25}$$

where the second-order terms like $\psi^{(2)}(\mathbf{x},t)$ also contain a stationary part due to the quadratic terms of wave components with itself. The first terms in the perturbation series are time independent, so

$$\overline{\eta} = \frac{1}{2}(\nabla\overline{\varphi}\cdot\nabla\overline{\varphi} - U^2)$$

is the stationary wave height. We consider small steady velocities, hence they are considered to be small enough to linearise the free surface as we described. Substituting in the Taylor series for the pressure at the actual hull surface and collecting equal powers of ε, we get

$$\begin{aligned}
\overline{p}_{\tilde{S}} &= \overline{p},\\
p_{\tilde{S}}^{(1)} &= p^{(1)} + \{\mathbf{X}^{(1)} + \mathbf{\Omega}^{(1)} \times (\overline{\mathbf{x}} - \mathbf{x}_g)\}\cdot\nabla\overline{p},\\
p_{\tilde{S}}^{(2)} &= p^{(2)} + \{\mathbf{X}^{(2)} + \mathbf{\Omega}^{(2)} \times (\overline{\mathbf{x}} - \mathbf{x}_g) + \mathbf{\Omega}^{(1)} \times [\mathbf{\Omega}^{(1)} \times (\overline{\mathbf{x}} - \mathbf{x}_g)]\}\cdot\nabla\overline{p}\\
&\quad + \{\mathbf{X}^{(1)} + \mathbf{\Omega}^{(1)} \times (\overline{\mathbf{x}} - \mathbf{x}_g)\}\cdot\nabla p^{(1)} + \frac{1}{2}\{[\mathbf{X}^{(1)} + \mathbf{\Omega}^{(1)} \times (\overline{\mathbf{x}} - \mathbf{x}_g)]\cdot\nabla\}^2\overline{p}.
\end{aligned} \tag{4.26}$$

From Bernoulli and the perturbation series for the potentials we get for the components of the pressure on the mean wetted surface:

$$\begin{aligned}
\overline{p} &= -\rho\left(gy_0 + \frac{1}{2}(\nabla\overline{\varphi}\cdot\nabla\overline{\varphi} - U^2)\right),\\
p^{(1)} &= -\rho\left(\nabla\overline{\varphi}\cdot\nabla\psi^{(1)} + \frac{\partial\psi^{(1)}}{\partial t}\right),\\
p^{(2)} &= -\rho\left(\frac{\partial\psi^{(2)}}{\partial t} + \frac{1}{2}\nabla\psi^{(1)}\cdot\nabla\psi^{(1)} + \nabla\psi^{(2)}\cdot\nabla\overline{\varphi}\right).
\end{aligned} \tag{4.27}$$

We carry out the pressure integration along the hull and apply high frequency filtering in the case of the multi frequency case and averaging in the case of one harmonic wave to obtain the second-order low frequency or constant drift force. The second-order force becomes

$$\begin{aligned}\mathbf{F}^{(2)} &= \frac{1}{2}\rho g \int_{WL} (\eta^{(1)} - \alpha_2^{(1)})^2 \overline{\mathbf{n}}\, \mathrm{d}l + \boldsymbol{\Omega}^{(1)} \times M \frac{\mathrm{d}^2 \mathbf{X}^{(1)}}{\mathrm{d}t^2} \\ &\quad - \rho \int_S \left\{ (\boldsymbol{\alpha}^{(1)} \cdot \nabla)\left(\frac{\partial \psi^{(1)}}{\partial t} + \nabla \psi^{(1)} \cdot \nabla \overline{\varphi} \right) + \frac{1}{2} \nabla \psi^{(1)} \cdot \nabla \psi^{(1)} \right. \\ &\quad \left. + \frac{\partial \psi^{(2)}}{\partial t} + \nabla \psi^{(2)} \cdot \nabla \overline{\varphi} \right\} \overline{\mathbf{n}}\, \mathrm{d}S. \end{aligned} \tag{4.28}$$

For the second-order moment we obtain a similar expression

$$\begin{aligned}\mathbf{M}^{(2)} &= \frac{1}{2}\rho g \int_{WL} (\eta^{(1)} - \alpha_2^{(1)})^2 (\overline{\mathbf{x}} - \mathbf{x}_g) \times \overline{\mathbf{n}}\, \mathrm{d}l + \boldsymbol{\Omega}^{(1)} \times M \frac{\mathrm{d}^2 \boldsymbol{\Omega}^{(1)}}{\mathrm{d}t^2} \\ &\quad - \rho \int_S \left\{ (\boldsymbol{\alpha}^{(1)} \cdot \nabla)\left(\frac{\partial \psi^{(1)}}{\partial t} + \nabla \psi^{(1)} \cdot \nabla \overline{\varphi} \right) + \frac{1}{2} \nabla \psi^{(1)} \cdot \nabla \psi^{(1)} \right. \\ &\quad \left. + \frac{\partial \psi^{(2)}}{\partial t} + \nabla \psi^{(2)} \cdot \nabla \overline{\varphi} \right\} (\overline{\mathbf{x}} - \mathbf{x}_g) \times \overline{\mathbf{n}}\, \mathrm{d}S. \end{aligned} \tag{4.29}$$

The influence of the second-order potentials in both the second-order forces and moments is considerable if one considers finite depth. Their influence on the second-order constant drift forces is negligible as follows from the analysis based on the far-field expansions, while in the computations of the constant second-order moments they may not be neglected. Many available computer codes take care of these effects in an approximate way see.

4.2.2 Constant Drift Forces by Means of Far-Field Expansions

Here we are mainly interested in the constant component of the drift force. In this section we apply a method that leads to results that are more accurate numerically. The direct pressure integration needs velocities obtained by differentiation of the potentials, depending on the numerical method to obtain the source strengths. This may lead to results that are not accurate enough for our purpose. Our choice is not to go to higher-order panel methods in the solver, but to use the results for the potentials and to avoid differentiation by applying the conservation of impulse for the fluid domain. This method is the one that in the past led to the first results of the drift forces. Here we restrict ourself to the determination of the mean drift forces.

The components of the horizontal mean drift forces, $\overline{F}_x$ and $\overline{F}_z$, and the moment around the vertical axis, $\overline{M}_y$, may be expressed as

$$\overline{F}_x = -\overline{\iint_{S_\infty} [p \cos\theta + \rho V_R (V_R \cos\theta - V_\theta \sin\theta)] R\, \mathrm{d}\theta\, \mathrm{d}y}, \tag{4.30}$$

$$\overline{F}_z = -\overline{\iint_{S_\infty} [p \sin\theta + \rho V_R (V_R \sin\theta + V_\theta \cos\theta)] R \, d\theta \, dy}, \tag{4.31}$$

$$\overline{M}_y = -\rho \overline{\iint_{S_\infty} V_R V_\theta R^2 \, d\theta \, dy}, \tag{4.32}$$

where p is the first-order hydrodynamic pressure, $\mathbf{V}$ is the fluid velocity with radial and tangential components V_R, V_θ and S_∞ is a large cylindrical control surface with radius R in the ship-fixed coordinate system. We derive from these formulas expressions in terms of the source densities of the first-order potentials

$$\sigma = \sigma^{(7)} + \sum_{j=1}^{6} \sigma^{(j)} \overline{\alpha}_j$$

where $\alpha_j = \overline{\alpha}_j e^{-i\omega t}$, $j = 1(1)6$ are the six modes of motion and the superscript 7 refers to the diffracted component of the source strength. However in our case, the velocity potential has the same form as before:

$$\begin{aligned} \varphi(\mathbf{x}, t) &= Ux + \overline{\phi}(\mathbf{x}; U) + \phi(\mathbf{x}; U) e^{-i\omega t} \\ &= Ux + \overline{\phi}(\mathbf{x}; U) + \left\{ \phi^{inc}(\mathbf{x}; U) + \phi^{(7)}(\mathbf{x}; U) + \sum_{j=1}^{6} \phi^{(j)}(\mathbf{x}; U) \overline{\alpha}_j \right\} e^{-i\omega t} \end{aligned} \tag{4.33}$$

where the potentials $\phi^{(j)}(\mathbf{x}; U)$, $j = 1, 7$ have the form (3.64) and are the potentials due to the motions and the diffraction. We assume that the potentials and the source strengths are expressed in terms of perturbation series (3.66), (3.67), and that the first two terms are known at this stage.

In the far field $R \gg 1$ we neglect the influence of the stationary potential $\overline{\phi}(\mathbf{x}; U)$ in (4.33), hence we approximate (4.33) by

$$\begin{aligned} \varphi(\mathbf{x}, t) = Ux &+ \frac{g\eta_a}{\omega_0} \exp\{k_1(\beta) y + i[k_1(\beta)(x \cos\beta + z \sin\beta) - i\omega t]\} \\ &+ F(\theta; U) e^{iS(\theta; U)} \sqrt{\frac{1}{R}} \exp\{k_1(\theta) y + i[k_1(\theta) R - i\omega t]\}; \end{aligned} \tag{4.34}$$

here η_a is the amplitude of the incoming wave in the direction β and the wave number $k_1(\beta)$:

$$\begin{aligned} k_1(\beta) &= \frac{g + 2\omega U \cos\beta - g\sqrt{1 + 4\tau \cos\beta}}{2U^2 \cos^2\beta} \\ &\approx \tilde{k}(1 - 2\tau \cos\beta) + O(\tau^2) = k_0 + O(\tau^2) \end{aligned} \tag{4.35}$$

where we use the notation $\tilde{k} = \omega^2/g$. Notice that this wave number is defined in the ship-fixed coordinate system and is different from $k_0 = \omega_0^2/g$, as defined before in the earth-fixed coordinate system. The function $F(\theta)e^{iS(\theta)}$ results from the expansion of the far-field potentials in (3.64). We obtain for large values of $|\mathbf{x}|$ and small values of U,

$$4\pi\phi^{(j)}(\mathbf{x};U) = \iint_S \sigma^{(j)}(\boldsymbol{\xi})\Psi(\mathbf{x},\boldsymbol{\xi})\,dS_\xi - \frac{2i\omega}{g}\iint_{FS} \nabla\overline{\phi}\cdot\nabla\phi^{(j)}\Psi(\mathbf{x},\boldsymbol{\xi})\,dS_\xi \tag{4.36}$$

where $\Psi(\mathbf{x},\boldsymbol{\xi})$ is the asymptotic expansion of the Green's function in the far field:

$$\Psi(\mathbf{x},\boldsymbol{\xi}) \approx 2\pi i\sqrt{\frac{2}{\pi R}}\frac{\exp\{i[k_1(\theta)R - \frac{\pi}{4}]\}}{[1+\frac{2U}{g}(\omega - k_1(\theta)U\cos\theta)\cos\theta]} \cdot \sqrt{k_1(\theta)}\exp\{k_1(\theta)(y+\eta) - ik_1(\theta)(\xi\cos\theta + \zeta\sin\theta)\} \tag{4.37}$$

with the local $k_1(\theta)$ wave number defined as

$$k_1(\theta) = \frac{g + 2\omega U\cos\theta - g\sqrt{1+4\tau\cos\theta}}{2U^2\cos^2\theta}. \tag{4.38}$$

Due to the fact that the function $\nabla\overline{\phi}(\boldsymbol{\xi})$ decays rapidly as $|\boldsymbol{\xi}| \to \infty$, it can be shown that the last term in (4.36) is approximated correctly with $\nabla\phi^{(j)}$ replaced by $\nabla\phi_0^{(j)}$. This leads to

$$\begin{aligned} F(\theta;U)e^{iS(\theta;U)} &= \sqrt{\frac{k_1(\theta)}{2\pi}}\left\{\frac{e^{\frac{\pi i}{4}}}{1+\frac{2U}{g}(\omega - k_1(\theta)U\cos\theta)\cos\theta}\right\} \\ &\cdot\left\{\iint_S(\sigma_0(\boldsymbol{\xi}) + \tau\sigma_1(\boldsymbol{\xi}) + \cdots)\exp\{k_1(\theta)\eta - ik_1(\theta)(\xi\cos\theta + \zeta\sin\theta)\}\,dS_\xi \right. \\ &\left. - 2i\tau\iint_{FS}\frac{\nabla\overline{\phi}}{U}\cdot\nabla\phi_0^{(T)}\exp\{-ik_1(\theta)(\xi\cos\theta + \zeta\sin\theta)\}\,dS_\xi\right\} \end{aligned} \tag{4.39}$$

where

$$\phi_0^{(T)} = \phi_0^{(7)} + \sum_{j=1}^{6}\phi_0^{(j)}\overline{\alpha}_j.$$

It is obvious that (4.39) can be written in the form

$$F(\theta;U)e^{iS(\theta;U)} = (1 - 2\tau\cos\theta)F_0(\theta)e^{iS_0(\theta)} + \tau F_1(\theta)e^{iS_1(\theta)} + O(\tau^2). \tag{4.40}$$

The functions $F_i(\theta)$ and $S_i(\theta)$ contain the local wave number $k_1(\theta)$.

The upper boundary of integration in (4.30) and (4.31) is the free surface

$$\tilde{\eta} = \frac{1}{g}\Re[(i\omega\phi(x,0,z) - U\phi_x(x,0,z))e^{-i\omega t}].$$

It follows from the pressure term that we can write

$$\overline{\int_{-\infty}^{\tilde{\eta}} p\,\mathrm{d}y} = \overline{\frac{\rho g}{2}\tilde{\eta}^2} - \frac{\rho}{2}\overline{\int_{-\infty}^{0} (|\mathbf{V}|^2 - U^2)\,\mathrm{d}y}.$$

We find the following expression for $\overline{F}_x$:

$$\begin{aligned}\overline{F}_x = &-\frac{\rho}{4}\int_0^{2\pi} \tilde{k}\phi\phi^* R\cos\theta\,\mathrm{d}\theta + \frac{\rho}{2}\tau\int_0^{2\pi} \Im(\phi\phi_x^*\cos\theta + \phi\phi_z^*\sin\theta)R\,\mathrm{d}\theta \\ &+ \frac{\rho}{4}\int_0^{2\pi}\int_{-\infty}^{0}\left[\left(\frac{1}{R^2}\phi_\theta\phi_\theta^* - \phi_R\phi_R^* + \phi_y\phi_y^*\right)R\cos\theta \right. \\ &\left. + (\phi_R\phi_\theta^* + \phi_\theta\phi_R^*)\sin\theta\right]\mathrm{d}\theta\,\mathrm{d}y \end{aligned} \tag{4.41}$$

and for $\overline{F}_z$:

$$\begin{aligned}\overline{F}_z = &-\frac{\rho}{4}\int_0^{2\pi} \tilde{k}\phi\phi^* R\sin\theta\,\mathrm{d}\theta - \frac{\rho}{2}\tau\int_0^{2\pi} \Im(\phi\phi_x^*\sin\theta - \phi\phi_z^*\cos\theta)R\,\mathrm{d}\theta \\ &+ \frac{\rho}{4}\int_0^{2\pi}\int_{-\infty}^{0}\left[\left(\frac{1}{R^2}\phi_\theta\phi_\theta^* - \phi_R\phi_R^* + \phi_y\phi_y^*\right)R\sin\theta \right. \\ &\left. - (\phi_R\phi_\theta^* + \phi_\theta\phi_R^*)\cos\theta\right]\mathrm{d}\theta\,\mathrm{d}y. \end{aligned} \tag{4.42}$$

In these expressions the asterisks denote the complex conjugate. The integration with respect to y needs some extra attention due to the fact that the exponential behaviour of the incident wave and the radiated or diffracted wave is different, due to the dependence of the wave number on β and θ, respectively. The function ϕ follows from (4.34):

$$\begin{aligned}\phi = &\frac{g\eta_a}{\omega_0}\exp\{k_1(\beta)y + \mathrm{i}k_1(\beta)(x\cos\beta + z\sin\beta)\} \\ &+ F(\theta;U)\mathrm{e}^{\mathrm{i}S(\theta;U)}\sqrt{\frac{1}{R}}\exp\{k_1(\theta)y + \mathrm{i}k_1(\theta)R\}. \end{aligned} \tag{4.43}$$

A closer look at (4.43) and (4.41) shows that the contributions to $\overline{F}_x$ consist of two parts. The first one, $F_x^{(1)}$, originates from those parts of the cross products that behave like $R^{-1/2}$ while the second one, $F_x^{(2)}$, originates from those square terms in (4.41) that behave like R^{-1}. We formally write:

$$\overline{F}_x = F_x^{(1)} + F_x^{(2)}. \tag{4.44}$$

After some lengthy manipulations and asymptotic expansion for large values of R we obtain for the mean surge force $F_x^{(1)}$ and the mean sway force $F_z^{(1)}$:

$$F_x^{(1)} \approx \mathscr{A}\sqrt{\frac{2\pi}{\tilde{k}}}F(\beta^*;U)\cos\left(S(\beta^*;U) + \frac{\pi}{4}\right)\cos\beta + \mathrm{O}(\tau^2) \tag{4.45}$$

and

$$F_z^{(1)} \approx \mathscr{A}\sqrt{\frac{2\pi}{\tilde{k}}}F(\beta^*;U)\cos\left(S(\beta^*;U)+\frac{\pi}{4}\right)\sin\beta+\mathrm{O}(\tau^2) \tag{4.46}$$

where

$$\mathscr{A} = -\frac{\rho\omega^2}{2\omega_0}\eta_a \quad \text{and} \quad \beta^* = \beta - 2\tau\sin\beta.$$

The second part of the wave-drift force may be analysed in the same way. We obtain

$$F_x^{(2)} \approx -\frac{\rho}{4}\tilde{k}\int_0^{2\pi} F^2(\theta;U)\{\cos\theta - 2\tau\sin^2\theta\}\,\mathrm{d}\theta + \mathrm{O}(\tau^2) \tag{4.47}$$

and

$$F_z^{(2)} \approx -\frac{\rho}{4}\tilde{k}\int_0^{2\pi} F^2(\theta;U)\{\sin\theta(1+2\tau\cos\theta)\}\,\mathrm{d}\theta + \mathrm{O}(\tau^2).$$

For the zero speed case this result is in accordance with the classical formula of Maruo [9] and for the general situation with Nossen et al. [13] if we change some signs due to the fact that our ship is sailing to the left. The formulation of the second-order moment acting on a vessel with constant forward speed and waves can be derived in a similar way [3].

4.3 Demonstration of Second-Order Effects, a Classroom Example

4.3.1 Interaction of Waves with a Vertical Wall

In this section we demonstrate the influence of reflection on the second-order wave effects in the two-dimensional case, of two plane waves acting on a vertical wall at $x = 0$. The wall is free to move within restraints by means of a linear spring. This case is chosen because it is possible to determine most of the essential parts of the first- and second-order effects explicitly.

We continue with the complex-valued potential function. The complex incident wave potential consists of the three components

$$\begin{aligned}\varphi_{inc} &= \alpha_1\frac{\cosh(k_1(y+h))}{\cosh(k_1h)}\mathrm{e}^{\mathrm{i}(\omega_1t-k_1x)} + \alpha_2\frac{\cosh(k_2(y+h))}{\cosh(k_2h)}\mathrm{e}^{\mathrm{i}(\omega_2t-k_2x+\vartheta)}\\ &\quad + \mathscr{C}\alpha_1\alpha_2\cosh(\delta k(y+h))\mathrm{e}^{\mathrm{i}(\delta\omega t-\delta kx+\vartheta)}\\ &= \varphi_{inc}^{(1)} + \varphi_{inc}^{(2)},\end{aligned} \tag{4.48}$$

where $\mathscr{C}$ is defined in (4.19).

At first we assume the wall to be fixed in space and that at the wall the normal velocity equals zero.

$$\frac{\partial \varphi}{\partial x} = 0 \quad \text{at } x = 0^-. \tag{4.49}$$

The first-order wave reflection is described by

$$\varphi^{(1)}_{refl} = \alpha_1 \frac{\cosh(k_1(y+h))}{\cosh(k_1 h)} \mathrm{e}^{\mathrm{i}(\omega_1 t + k_1 x)} + \alpha_2 \frac{\cosh(k_2(y+h))}{\cosh(k_2 h)} \mathrm{e}^{\mathrm{i}(\omega_2 t + k_2 x + \vartheta)} \tag{4.50}$$

because the reflection coefficients are equal to 1. The corresponding bound second-order, slowly varying, wave contribution becomes

$$\varphi^{(2)}_{refl} = \mathscr{C} \alpha_1 \alpha_2 \cosh(\delta k(y+h)) \mathrm{e}^{\mathrm{i}(\delta\omega t + \delta k x + \vartheta)}. \tag{4.51}$$

This second-order, slowly varying, potential is the *bound* wave for the reflected wave if we take the first-order reflected waves into account only. However in the free surface condition for the second-order potential (4.7) quadratic terms due to the product of the incident potential and reflected potential have to be taken into account.

It is convenient to combine the first-order terms first and then to compute its contribution to the total slowly varying second-order potential. Hence we consider the second-order contribution of

$$\begin{aligned} \varphi^{(1)}(x, y, t) &= \Re\left(\varphi^{(1)}_{inc} + \varphi^{(1)}_{refl}\right) \\ &= 2 \sum_{j=1,2} \alpha_j \frac{\cosh(k_j(y+h))}{\cosh(k_j)h} \cos(k_j x)\cos(\omega_j t + \vartheta_j) \end{aligned} \tag{4.52}$$

where $\vartheta_1 = 0$ and $\vartheta_2 = \vartheta$. The free surface condition for the second-order slowly varying potential becomes

$$\begin{aligned} g\varphi^{(2)}_y + \varphi^{(2)}_{tt} \approx\ & 2\alpha_1\alpha_2\delta\omega \sin(\delta\omega t + \vartheta) \\ & \cdot \Big\{ -k^2 \sin(k_1 x)\sin(k_2 x) \\ & + \left(\frac{k^2}{2} - \frac{k\omega}{c_g} - \frac{\omega^4}{2g^2}\right) \cos(k_1 x)\cos(k_2 x) \Big\}, \end{aligned} \tag{4.53}$$

where we have made use of the dispersion relation for the first-order waves, to simplify equation (4.53) can be rewritten in the form

$$\begin{aligned} g\varphi^{(2)}_y + \varphi^{(2)}_{tt} &\approx \alpha_1\alpha_2 \Big\{ -\left(\frac{1}{2}k^2 + \frac{k\omega}{c_g} + \frac{\omega^4}{2g^2}\right)\cos(\delta k x) \\ &\quad + \left(\frac{3}{2}k^2 - \frac{k\omega}{c_g} - \frac{\omega^4}{2g^2}\right)\cos((k_1+k_2)x) \Big\} \delta\omega \sin(\delta\omega t + \vartheta) \\ &= \left\{ \mathscr{C}_1 \cos(\delta k x) + \mathscr{C}_2 \cos((k_1+k_2)x) \right\} \delta\omega \sin(\delta\omega t + \vartheta), \end{aligned} \tag{4.54}$$

with

$$\mathscr{C}_1 = -\alpha_1\alpha_2\left(\frac{1}{2}k^2 + \frac{k\omega}{c_g} + \frac{\omega^4}{2g^2}\right),$$
$$\mathscr{C}_2 = \alpha_1\alpha_2\left(\frac{3}{2}k^2 - \frac{k\omega}{c_g} - \frac{\omega^4}{2g^2}\right).$$

Hence the slowly varying potential function consists of a term with *slow* δk variations in the x-direction and *fast* variations $k_1 + k_2$. The latter term is the interaction effect of the incident and reflected waves. The solution can be obtained by means of separation of variables:

$$\varphi^{(2)}(x, y, t) = \left\{\frac{\mathscr{C}_1 \cos(\delta k x)\cosh(\delta k(y+h))}{g\delta k \sinh(\delta k h) - (\delta\omega)^2\cosh(\delta k h)} + \frac{\mathscr{C}_2 \cos((k_1+k_2)x)\cosh((k_1+k_2)(y+h))}{g(k_1+k_2)\sinh((k_1+k_2)h)}\right\}\delta\omega\sin(\delta\omega t + \vartheta). \tag{4.55}$$

The combined bound, slowly varying, wave contributions obey the boundary condition at $x = 0$. This is only valid in this simple case. The second part of (4.55) is asymptotically small compared with the first part, hence we may ignore it. The final result thus equals asymptotically the sum of (4.18) and (4.51). However, we do not ignore the small term to determine its effect on the second-order force.

For more complex geometrical objects one also has to deal with a free second-order contribution to fulfil the boundary condition at the body surface.

In the next section it will be shown that, if the wall is free to move, the corresponding reaction potential also plays a role in the free surface condition (4.11).

4.3.2 Forces on a Fixed Wall

We use the same problem of the fixed impenetrable plate to demonstrate the way the second-order, slowly varying, horizontal force may be calculated. The pressure at the plate is given by (1.13)

$$\varphi_t + \frac{1}{2}(\varphi_x^2 + \varphi_y^2) + gy + \frac{p - p_0}{\rho} = 0. \tag{4.56}$$

The time dependent horizontal force, per unit length, acting on the wall equals

$$F_x(t) = \int_{-h}^{\eta(0,t)} p(0, y, t)\,\mathrm{d}y := F^{(1)}(t) + F^{(2)}(t) + \cdots, \tag{4.57}$$

where $F^{(1)}(t)$ is the first-order force and $F^{(2)}(t)$ the slowly varying force. We ignore the other second-order terms. The first-order force equals

$$F^{(1)}(t) = -\rho \int_{-h}^{0} \varphi_t^{(1)}(0, y, t)\,\mathrm{d}y = -2\mathrm{i}\rho \sum_{j=1,2} \frac{\alpha_j \omega_j^3}{g k_j^2} \mathrm{e}^{\mathrm{i}(\omega_j t + \vartheta_j)}. \tag{4.58}$$

The low frequency second-order force consists of a contribution of the second-order, slowly varying, potential and of quadratic first-order terms. If the plate is free to move we also have to take into account the effect of the flow field of the motion and the effect of the local displacement of the plate. However, our plate is fixed in space. The force due to the second-order, slowly varying, potential (4.55) becomes

$$F^{(2a)}(t) = \rho \left\{ \frac{\mathscr{C}_1 c_g \delta\omega \sinh(\delta k h)}{g\delta k \sinh(\delta k h) - (\delta\omega)^2 \cosh(\delta k h)} + \frac{\mathscr{C}_2 (\delta\omega)^2 \sinh(2kh)}{4gk^2 \sinh(2kh)} \right\} \cdot \cos(\delta\omega t + \vartheta). \tag{4.59}$$

The second part in expression (4.59) is small $\mathrm{O}((\delta\omega)^2)$ compared with the first term $\mathrm{O}(1)$, hence we may ignore it. The part generated by the first-order potentials equals

$$F^{(2b)}(t) = \frac{\rho}{2} \left\{ \frac{1}{g} \varphi_t^2 \Big|_{y=0} - \int_{-h}^{0} \varphi_y^2\,\mathrm{d}y \right\}, \tag{4.60}$$

because $\varphi_x = 0$ at $x = 0$. If we insert (4.52) into (4.60) we obtain for the slowly varying part

$$F^{(2b)}(t) = \frac{\rho}{2} \alpha_1 \alpha_2 \left\{ \frac{\omega^2}{g} - \frac{k^2 \sinh(\delta k h)}{\delta k \cosh^2(kh)} \right\} \cos(\delta\omega t + \vartheta). \tag{4.61}$$

Both terms in (4.61) are of the same order of magnitude as the first term in (4.59). In the case of a fixed wall these two terms are the only contribution to the second-order slowly varying force.

4.3.3 Moving Wall

If the wall is able to move the situation becomes more complicated. We restrict ourselves to the case that the wall is free to translate in horizontal direction. We consider the situation shown in Fig. 4.1.

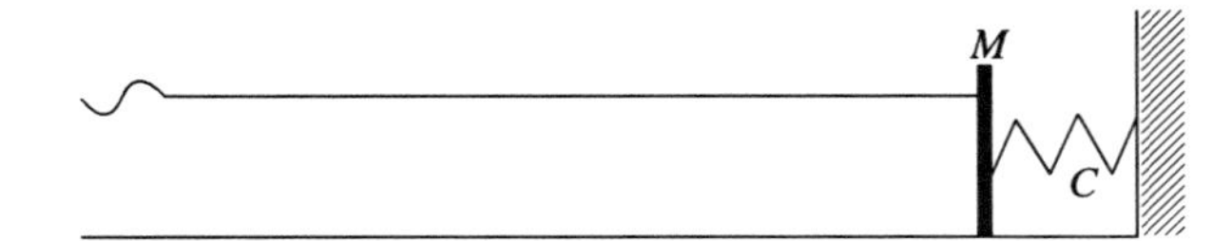

Fig. 4.1 Moving wall

The wall has a mass M and the motion is restrained by means of a linear spring with spring constant C. Other configurations can be treated in the same way. The motion of the wall consists of a response to the first-order force with the frequencies of the incident first-order waves and second-order components. The first-order motion will generate a reaction force in and out of phase with the motion. The force in phase with the motion leads to the added mass, while the out of phase part gives the wave damping. The second-order low frequency force now consist of several terms. First of all the forces described in (4.59) and (4.61) act on the wall, but also the effect of the 'high' frequency displacement of the wall gives a contribution that can not be ignored. So first we study the high frequency first-order motion.

4.3.4 First-Order Motion of the Wall

In the case that the wall is free to translate in the horizontal direction only, the equation of motion can be written as

$$M\frac{\mathrm{d}^2\bar{x}}{\mathrm{d}t^2} + C\bar{x} = F_{exc}(t) + F_{react}(t), \tag{4.62}$$

where $F_{exc}(t)$ is the excitation force due to the incident waves and $F_{react}(t)$ the reaction force due to the motion. In general the time dependent reaction force can be written as

$$F_{react}(t) = a\frac{\mathrm{d}^2\bar{x}}{\mathrm{d}t^2}(t) + \int_{-\infty}^{t} \mathscr{L}(t-\tau)\frac{\mathrm{d}\bar{x}}{\mathrm{d}\tau}(\tau)\,\mathrm{d}\tau.$$

Because we consider the linear part of the problem we may consider the equation of motion for each frequency separately. We consider incident waves with frequency ω, hence we write $\bar{x} = x_w \mathrm{e}^{\mathrm{i}\omega t}$. The excitation force equals (4.58) for a single frequency ω and the reaction force can be rewritten as

$$F_{react}(t) = F_{react}\mathrm{e}^{\mathrm{i}\omega t} = (\omega^2 a(\omega) - \mathrm{i}\omega b(\omega))x_w \mathrm{e}^{\mathrm{i}\omega t}. \tag{4.63}$$

Analogously with the theory of ship motions we write the equation of motion as follows

$$(M + a(\omega))\frac{\mathrm{d}^2\bar{x}}{\mathrm{d}t^2} + b(\omega)\frac{\mathrm{d}\bar{x}}{\mathrm{d}t} + C\bar{x} = F_{exc}(t), \tag{4.64}$$

where $a(\omega)$ is the frequency dependent added mass and $b(\omega)$ the frequency dependent wave damping coefficient. If the wall is forced to move with amplitude x_w, the boundary value problem for the two-dimensional linearised wave potential

$\bar{\varphi}(x, y, t) = \varphi(x, y)e^{i\omega t}$ becomes

$$\begin{aligned}
&\varphi_{xx} + \varphi_{yy} = 0, \\
&\varphi_y = \frac{\omega^2}{g}\varphi \quad \text{at } y = 0, \\
&\varphi_y = 0 \quad \text{at } y = -h, \\
&\varphi_x = i\omega x_w \quad \text{at } x = 0
\end{aligned} \tag{4.65}$$

furthermore we permit in the far field only wave travelling away from the wall. The solution can be written as a sum of eigenfunctions as described before,

$$\varphi(x, y) = \sum_{m=0}^{\infty} a_m \cosh(k_m(y + h))e^{ik_m x}, \tag{4.66}$$

where k_0 is the positive real root of the dispersion relation $k \sinh(kh) = \frac{\omega^2}{g} \cosh(kh)$ and $k_m = -i\kappa_m$, for $m = 1, 2, \ldots$, are the negative imaginary roots of the same relation. If we use the orthogonality relation the boundary condition at $x = 0$ yields

$$a_m = \frac{4\omega \sinh(k_m h)}{k_m\{\sinh(2k_m h) + 2k_m h\}} x_w. \tag{4.67}$$

The first-order force on the wall due to the motion becomes

$$\begin{aligned}
F_{react} &= -i\omega\rho \int_{-h}^{0} \varphi(0, y)\, dy \\
&= -4i\omega^2\rho \sum_{m=0}^{\infty} \frac{\sinh^2(k_m h)}{k_m^2\{\sinh(2k_m h) + 2k_m h\}} x_w.
\end{aligned} \tag{4.68}$$

The added mass and wave damping becomes

$$\begin{aligned}
a(\omega) &= 4\rho \sum_{m=1}^{\infty} \frac{\sin^2(\kappa_m h)}{\kappa_m^2\{\sin(2\kappa_m h) + 2\kappa_m h\}}, \\
b(\omega) &= 4\omega\rho \frac{\sinh^2(k_0 h)}{k_0^2\{\sinh(2k_0 h) + 2k_0 h\}}.
\end{aligned} \tag{4.69}$$

If, for the one wave system, we combine (4.58) with (4.64) we obtain for the amplitude of the motion of the wall

$$x_w = -2i\rho \frac{\alpha\omega^3}{g k_0^2\{-(M + a(\omega))\omega^2 + ib(\omega)\omega + C\}}. \tag{4.70}$$

For the two-wave system we may write for the first-order motion

$$\bar{x}^{(1)}(t) = -2\mathrm{i}\rho \sum_{j=1,2} \frac{\alpha_j \omega_j^3 \mathrm{e}^{\mathrm{i}(\omega_j t + \vartheta_j)}}{g k_{0j}^2 \{-(M + a(\omega_j))\omega_j^2 + \mathrm{i}b(\omega_j)\omega_j + C\}}. \tag{4.71}$$

If we now reconsider the free surface condition for the slowly varying potential function (4.11), we see that the right-hand side consists of terms from the incident and reflected wave and reaction wave. Generally these effects are hardly taken into account in numerical codes. In this simple case we can determine this potential analytically. The wave travelling to the left now consists of the reflected wave and the reaction wave, so instead of (4.50) we have the following travelling wave to the left $\varphi_{refl} + \varphi_{react}$ with

$$\varphi_{react}^{(1)} = \beta_1 \frac{\cosh(k_1(y+h))}{\cosh(k_1 h)} \mathrm{e}^{\mathrm{i}(\omega_1 t + k_1 x + \theta_1)} + \beta_2 \frac{\cosh(k_2(y+h))}{\cosh(k_2 h)} \mathrm{e}^{\mathrm{i}(\omega_2 t + k_2 x + \theta_2)}, \tag{4.72}$$

where $k_{1,2}$ are the real roots of the dispersion relation and

$$\beta_{1,2} \mathrm{e}^{\mathrm{i}\theta_{1,2}} = \frac{2\omega_{1,2} \sinh(2k_{1,2}h)}{k_{1,2}(\sinh(2k_{1,2}h) + 2k_{1,2}h)} x_w^{1,2}.$$

It can be proven that the effect of the imaginary roots of the dispersion relation on the second-order potential becomes asymptotically small compared with the effect of the travelling wave. Hence we ignore the effect of the evanescent wave. The second-order potential bound to the potential in (4.72) equals expression (4.51) with α replaced by β,

$$\varphi_{react}^{(2)} = \mathscr{C} \beta_1 \beta_2 \cosh(\delta k(y+h)) \mathrm{e}^{\mathrm{i}(\delta\omega t + \delta k x + \delta\theta)}. \tag{4.73}$$

This wave potential does not obey the boundary condition at $x = 0$, hence we have to consider a free wave $\varphi^{(2')}$ at the same frequency $\delta\omega$. We have as boundary condition for the free second-order slowly varying potential

$$\varphi_x^{(2')}|_{x=0} = -\mathscr{C} \beta_1 \beta_2 \delta k \cosh(\delta k(y+h)) \mathrm{e}^{\mathrm{i}(\delta\omega t + \delta\theta)}. \tag{4.74}$$

Next we write

$$\varphi^{(2')}(x, y, t) = -\mathscr{C} \beta_1 \beta_2 \delta k \psi(x, y) \mathrm{e}^{\mathrm{i}(\sigma t + \delta\theta)},$$

where $\sigma = \delta\omega$, and obtain for ψ the boundary value problem

$$\begin{aligned}
&\psi_{xx} + \psi_{yy} = 0, \\
&\psi_y - \frac{(\sigma)^2}{g}\psi = 0 && \text{at } y = 0, \\
&\psi_y = 0 && \text{at } y = -h, \\
&\psi_x = \cosh(\delta k(y+h)) = f(y) && \text{at } x = 0.
\end{aligned} \tag{4.75}$$

It should be noticed here δk is not a solution of the linear dispersion relation $k\sinh(kh) - \frac{\sigma^2}{g}\cosh(kh) = 0$. The zero's of the dispersion relation are the real values $k_0 = \pm\tilde{k}$ and the imaginary values $k_j = \pm\mathrm{i}\tilde{\kappa}_j$, $j = 1, 2, \ldots$. Because the waves are travelling in the direction of the negative x-axis we only consider the real positive $+k_0$ and the negative imaginary values $-\mathrm{i}\tilde{\kappa}_j$. The set of functions $\cosh(k_j(y+h))$ form a complete orthogonal set of functions. It is easy to show that

$$\int_{-h}^{0} \cosh(k_i(y+h))\cosh(k_j(y+h))\,\mathrm{d}y = 0 \quad \text{for } i \neq j.$$

Hence we write the solution as

$$\psi(x, y) = \sum_{j=0}^{\infty} a_j \cosh(k_j(y+h))\mathrm{e}^{\mathrm{i}k_j x}, \tag{4.76}$$

with

$$a_j = \frac{\int_{-h}^{0} f(y)\cosh(k_j(y+h))\,\mathrm{d}y}{\mathrm{i}k_j \int_{-h}^{0} \cosh^2(k_j(y+h))\,\mathrm{d}y}$$
$$= \frac{2\{\frac{\sinh((k_j+\delta k)h)}{k_j+\delta k} + \frac{\sinh((k_j-\delta k)h)}{k_j-\delta k}\}}{\mathrm{i}\{\sinh(2k_j h) + 2k_j h\}}. \tag{4.77}$$

It should be noticed that $\tilde{k}$ and δk are both small, thus we consider them to be of the same order of magnitude. So, we may conclude that the travelling part is dominant compared with the evanescent contributions. This leads to the following approximation of the free second-order wave

$$\varphi^{(2')}(x, y, t) = -\frac{2}{\mathrm{i}}\mathscr{C}\beta_1\beta_2\delta k\left\{\frac{\sinh((\tilde{k}+\delta k)h)}{\tilde{k}+\delta k} + \frac{\sinh((\tilde{k}-\delta k)h)}{\tilde{k}-\delta k}\right\}$$
$$\cdot \frac{\cosh(\tilde{k}(y+h))\mathrm{e}^{\mathrm{i}(\sigma t+\tilde{k}x+\delta\theta)}}{\{\sinh(2\tilde{k}h) + 2\tilde{k}h\}}$$
$$= \mathscr{C}\gamma_1\gamma_2\cosh(\tilde{k}(y+h))\mathrm{e}^{\mathrm{i}(\sigma t+\tilde{k}x+\delta\theta)}, \tag{4.78}$$

where

$$\gamma_j = \frac{-2\beta_j\delta k}{\mathrm{i}\{\sinh(2\tilde{k}h) + 2\tilde{k}h\}}\left\{\frac{\sinh((\tilde{k}+\delta k)h)}{\tilde{k}+\delta k} + \frac{\sinh((\tilde{k}-\delta k)h)}{\tilde{k}-\delta k}\right\}.$$

Furthermore the second-order slowly varying, free wave is of the same order of magnitude as the bound slowly varying waves.

4.3.5 Influence of the Motion on the Low Frequency Drift Force

If the wall is moving (4.58) contains a quadratic term as well because the potential has to be taken at the actual position. We rewrite (4.58) as

$$F^{(1')}(t) = -\rho \int_{-h}^{0} \varphi_t^{(1)}(\bar{x}(t), y, t)\,\mathrm{d}y \approx F^{(1)}(t) - \rho \bar{x}(t) \int_{-h}^{0} \varphi_{xt}^{(1)}(0, y, t)\,\mathrm{d}y. \tag{4.79}$$

Because in this special case $\varphi_x^{(1)}(0, y, t) = 0$ the influence of the first-order motion on the low frequency drift force equals zero. In the case of wave interaction with a moored object it may give a contribution of the same order of magnitude as the two terms (4.59) and (4.61).

4.3.6 Second-Order Motion of the Wall

The contribution to the second-order force of the *bound* and *free* second-order wave potentials due to the first-order motion of the wall equals

$$F^{(2c)}(t) = -\rho\delta\omega\Re\left\{\mathrm{i}\mathscr{C}\left(\frac{\beta_1\beta_2\sinh(\delta k h)}{\delta k} + \frac{\gamma_1\gamma_2\sinh(\tilde{k}h)}{\tilde{k}}\right)\mathrm{e}^{\mathrm{i}(\delta\omega t + \delta\theta)}\right\}. \tag{4.80}$$

This term is of the same order of magnitude compared with $F^{(2a)}(t)$ and $F^{(2b)}(t)$ because $\mathscr{C}\delta\omega = \mathrm{O}(1)$ for small values of $\delta\omega = \sigma$. The total low frequency second-order force acting on the wall equals

$$F^{(2)}(t) = F^{(2a)}(t) + F^{(2b)}(t) + F^{(2c)}(t) \doteqdot \Re\{\overline{F}^{(2)}\mathrm{e}^{\mathrm{i}\sigma t}\mathrm{e}^{\mathrm{i}\sigma t}\}. \tag{4.81}$$

The low frequency motion is described by (4.64)

$$(M + a(\sigma))\frac{\mathrm{d}^2\bar{x}}{\mathrm{d}t^2} + b(\sigma)\frac{\mathrm{d}\bar{x}}{\mathrm{d}t} + C\bar{x} = \Re\{\overline{F}^{(2)}\mathrm{e}^{\mathrm{i}\sigma t}\}. \tag{4.82}$$

The harmonic solution of this equation becomes

$$\bar{x}(t) = \Re\left\{\frac{\overline{F}^{(2)}\mathrm{e}^{\mathrm{i}\sigma t}}{-(M + a(\sigma))\sigma^2 + \mathrm{i}b(\sigma)\sigma + C}\right\}. \tag{4.83}$$

For a soft mass-spring system near resonance with $M\sigma^2 \approx C$ the amplitude of motion becomes very large for small values of σ. At this point we have reached a contradiction with our earlier made assumptions. We have to take into account the effect of large motions in the formulation of the first-order potential function. This in its turn influences the motion at the second-order low frequency. Hence we may assume that the right-hand side of the equation of motion (4.82) depends on the low frequency velocity of the object. Hence $\overline{F}^{(2)}$ is a function of $\frac{\mathrm{d}\bar{x}}{\mathrm{d}t} = U(\sigma t)$ and we assume that we may expand this dependency by means of a Taylor series that we cut of after the second term as follows:

$$\overline{F}^{(2)}(U) = \overline{F}^{(2)}(0) + U\frac{\mathrm{d}\overline{F}^{(2)}}{\mathrm{d}U}(0) + \mathrm{O}(U^2). \tag{4.84}$$

Hence we have to take into account an extra damping term equal to $\frac{\mathrm{d}\overline{F}^{(2)}}{\mathrm{d}U}(0)$ in the equation of motion. This damping is called the wave drift damping. We leave it as an open question how to compute this damping term. In general its value is much larger than the wave damping in the equation of motion. Besides this effect viscous damping may play a role in the value of the damping as well. This does not alter the conclusion that the amplitude of motion for a softly anchored wall remains significantly large.

4.3.7 Some Observations

This analysis indicates that if one considers a ship moored to a buoy it is mostly impossible to take all the effects into account. Some effects that could be computed in the wall case may be taken into account approximately. In general the damping terms caused by viscous effects are sufficient to take into account, however for high sea states the damping due to the second order motion may become relevant. One must keep in mind that the latter is quadratic with respect to the wave height and may become dominant in high sea states.

Chapter 5
Asymptotic Formulation

In this chapter we describe the flow past a thin ship with uniform speed. In Chap. 2 we derived the field of a steady source positioned in a uniform flow by means of Fourier transforms. Here the steady field around a thin ship will be written as a superposition of sources at the centre-plane of the ship. For the wave resistance the theory results in the classical *Michell integral* as published in (1898) for the first time.

The second part of this chapter is concerned with short wave diffraction by a blunt ship at constant forward speed. The speed of the ship is considered to be small. If the double body potential around the ship is known one can use an asymptotic short wave theory. The method used is the so called *ray method* which is well known in acoustic theory. We show some results for shapes where an expression for the double body potential is known explicitly, i.e. a circular cylinder and a sphere. In literature one may find some results for a ship shape.

5.1 Thin Ship Hydrodynamics, *Michell* Theory

We consider a symmetrical ship moving with a constant speed U in the direction of the $\bar{x}$-axis of the right-handed coordinate system $\bar{x}, \bar{y}, \bar{z}$ where the $\bar{y}$-axis is positive in upward (originally in the Michell theory downward) direction. For convenience, we shall fix the coordinate system to the ship, or in other words, we consider a steady flow past the ship. The incoming flow has a uniform speed U in the negative x-direction. We assume the equation of the hull to be given by $\bar{z} = \pm B\bar{f}(\bar{x}, \bar{y})$ where $\bar{f}$ is a smooth function defined on the projection of the ship hull on the $\bar{x}, \bar{y}$-plane. The beam B of the ship is small compared to its length L. The perturbation caused by the ship is described by a velocity potential $\bar{\varphi}$ which satisfies a Laplace equation together with the linearised free surface condition, obtained from (1.32) which now reads

$$U^2\bar{\varphi}_{\bar{x}\bar{x}} + g\bar{\varphi}_{\bar{y}} = 0, \quad \text{at } \bar{y} = 0 \tag{5.1}$$

A.J. Hermans, *Water Waves and Ship Hydrodynamics*,
DOI 10.1007/978-94-007-0096-3_5,

and the boundary condition on the ship hull. This boundary condition states that the total normal velocity relative to the ship vanishes. According to the coordinate system used, we have an unperturbed speed $-U$ in the x-direction and hence the velocity is $(-U + \bar{\varphi}_{\bar{x}}, \bar{\varphi}_{\bar{y}}, \bar{\varphi}_{\bar{z}})$. Since $(\pm \bar{f}_{\bar{x}}, \pm \bar{f}_{\bar{y}}, -1)$ are direction numbers of the normal to the ship hull, we obtain finally the condition

$$-U \bar{f}_{\bar{x}} + \bar{\varphi}_{\bar{x}} \bar{f}_{\bar{x}} + \bar{\varphi}_{\bar{y}} \bar{f}_{\bar{y}} \mp \bar{\varphi}_{\bar{z}} = 0. \tag{5.2}$$

Now we introduce dimensionless coordinates $x = \bar{x}/L, y = \bar{y}/L, z = \bar{z}/L$ and the dimensionless potential φ defined by

$$\bar{\varphi} = \varepsilon U L \phi(x, y, z),$$

with $\varepsilon = B/L$. Clearly this does not affect the Laplace equation. However, the free surface condition (5.1) becomes

$$F^2 \varphi_{xx} + \varphi_y = 0, \quad \text{at } y = 0, \tag{5.3}$$

where $F = U/\sqrt{gL}$ is the Froude number. Furthermore, the condition (5.2) is transformed in dimensionless form,

$$-f_x + \varepsilon \varphi_x f_x + \varepsilon \varphi_y f_y \mp \varphi_z = 0, \quad \text{at } z = \pm \varepsilon f(x, y), \tag{5.4}$$

by setting $\bar{f}(x, y) = B f(x, y)$. In the first-order approximation, this condition reduces to

$$\varphi_z = \mp f_x \tag{5.5}$$

on the plane $z = 0$. Due to the symmetry of the flow we see that outside the ship hull $\varphi_z = 0$, for $z = 0$. We are thus led to the boundary value problem for φ defined by the Laplace equation

$$\varphi_{xx} \varphi_{yy} + \varphi_{zz} = 0, \tag{5.6}$$

together with the conditions

$$\varphi \text{ finite}, \quad \text{as } y \to -\infty, \tag{5.7}$$

$$\varphi_z = \begin{Bmatrix} \mp f_x & \text{inside} \\ 0 & \text{outside} \end{Bmatrix} \quad \text{the projection of the ship hull on the plane } z = 0.$$

There are two ways of solving this problem. The first one uses the Fourier transforms; the other one is by means of a distribution of sources on the centre-plane. In Chap. 2 we derived, for deep water, an expression for the field of a unit source in a steady current by means of Fourier transforms (2.125), therefore we follow the second method in this section. We use a slightly different notation for this so-called Green's function

$$\mathscr{G}(x, y, z; \tilde{x}, \tilde{y}, \tilde{z}) = -\frac{1}{4\pi r} + \frac{1}{4\pi \bar{r}} - \frac{1}{2\pi^2} \int_0^\infty \mathrm{d}k \int_0^\pi \mathrm{d}\vartheta \frac{1}{1 - k\upsilon \cos^2 \vartheta} \cdot \mathrm{e}^{k((y+\tilde{y}) - \mathrm{i}(x-\tilde{x})\cos\vartheta)} \cos(k(z-\tilde{z})\sin\vartheta). \tag{5.8}$$

The solution of the boundary value problem is written as the following distribution of sources at the projection of the hull on the centre-plane $\mathscr{S}_0$ with strength $\sigma(\tilde{x}, \tilde{y})$:

$$\varphi(x, y, z; \tilde{x}, \tilde{y}, \tilde{z}) = \iint_{\mathscr{S}_0} \mathscr{G}(x, y, z; \tilde{x}, \tilde{y}, 0)\sigma(\tilde{x}, \tilde{y})\,\mathrm{d}\tilde{x}\,\mathrm{d}\tilde{y}. \tag{5.9}$$

We apply the condition on the centre-plane and make use of the symmetry of the Green's function and (3.17). This results in

$$\sigma(\tilde{x}, \tilde{y}) = -2f_x(\tilde{x}, \tilde{y}). \tag{5.10}$$

The resistance of the ship can be found by integrating the pressure distribution over the hull. The pressure distribution along the hull follows from the Bernoulli equation (1.13). Since the pressure is symmetric with respect to the centre-plane, we obtain for the component in The x-direction of the total force $\mathscr{R}$,

$$\mathscr{R} = 2\rho U \iint_{\mathscr{S}_0} \varphi_x(x, y, 0) f_x(x, y)\,\mathrm{d}x\,\mathrm{d}y. \tag{5.11}$$

Working out this expression we see that the term $1/r$ and $1/\bar{r}$ do not give any contribution to $\mathscr{R}$, as they should not do according to the d'Alembert paradox which states that the total force acting on an object located in a potential flow, without vorticity, is equal to zero.

After rather tedious computations, the final result is

$$\mathscr{R} = \frac{4g^2\rho}{\pi U^2} \int_0^{\frac{\pi}{2}} \sec^3\vartheta [P^2(\vartheta) + Q^2(\vartheta)]\,\mathrm{d}\vartheta, \tag{5.12}$$

where

$$P = \iint_{\mathscr{S}_0} f_x \mathrm{e}^{y\frac{\sec^2\vartheta}{F^2}} \cos\left(\frac{x\sec\vartheta}{F^2}\right) \mathrm{d}x\,\mathrm{d}y,$$

$$Q = \iint_{\mathscr{S}_0} f_x \mathrm{e}^{y\frac{\sec^2\vartheta}{F^2}} \sin\left(\frac{x\sec\vartheta}{F^2}\right) \mathrm{d}x\,\mathrm{d}y.$$

The results (5.12) may be put in a variety of different forms by change of variable and order of integration. If one poses $\lambda = \sec\vartheta$, then it can be verified that

$$\mathscr{R} = \frac{4g^2\rho}{\pi U^2} \iint_{\mathscr{S}_0} \mathrm{d}x\,\mathrm{d}y \iint_{\mathscr{S}_0} \mathrm{d}\tilde{x}\,\mathrm{d}\tilde{y}\, f_x(x, y) f_x(\tilde{x}, \tilde{y}) M\left(\frac{x-\tilde{x}}{F^2}, \frac{y-\tilde{y}}{F^2}\right), \tag{5.13}$$

where

$$M(x, y) = \int_1^{\infty} \frac{\lambda^2}{\sqrt{\lambda^2 - 1}} \mathrm{e}^{\lambda^2 y} \cos\lambda x\,\mathrm{d}\lambda.$$

Expression (5.13) for $\mathscr{R}$ is usually called 'Michell's integral' (1898). In practice this expression gives a reasonable approximation for a limited class of ships sailing at moderate values of the Froude number. For small values of F, for instance for very large vessels, it overestimates the wave resistance greatly, while for blunt ships it is not valid at all. It is generally understood nowadays that for these hulls the complete non-linear free surface problem must be solved numerically. In the chapter on numerical methods we shall again pay attention to the computation of the wave resistance.

5.2 Short Wave Diffraction by a Sailing Ship

In this section we describe an asymptotic short wave theory to compute the added resistance for a ship sailing slowly in deep water. In this approach the Froude number, defined with respect to the length of the ship, $Fn = U/\sqrt{gL}$ is assumed to be small as well. In this case the steady potential $\overline{\phi}_s(\mathbf{x})$ can be replaced by the double body potential $\overline{\phi}_r(\mathbf{x})$. In Chap. 1 we derived a free surface condition for the wave potential. In the coordinate system

$$x' = x, \qquad y' = y - \eta_r(x,z), \qquad z' = z,$$

we obtain a boundary condition for the unsteady wave potential $\phi(\mathbf{x},t)$ that, after omitting primes, becomes

$$\frac{1}{g}\left[\frac{\partial}{\partial t} + u\frac{\partial}{\partial x} + w\frac{\partial}{\partial z}\right]^2 \phi + \frac{\partial}{\partial y}\phi = 0 \quad \text{on } y = 0, \tag{5.14}$$

where η_r is the free surface elevation due to ϕ_r and the horizontal velocity $\mathbf{u} = (u,w) = \nabla_2\phi_r$ is calculated at the undisturbed free surface. The terms in this expression must be of the same order of magnitude. This is the case if the frequency of the waves is large, while the dimensionless parameter $\tau = \omega U/g$ remains finite. It can be shown that the neglected terms in the free surface condition are small in this situation.

The potential function $\phi(\mathbf{x},t)$ obeys the Laplace equation

$$\Delta\phi = 0 \quad \text{in the fluid} \tag{5.15}$$

and on the ship's hull we have $\frac{\partial\phi}{\partial n} = 0$. At infinity the incoming wave field consists of a plane wave

$$\phi_{inc} = \mathrm{e}^{\mathrm{i}k_0(x\cos\theta + z\sin\theta) + k_0 y + \mathrm{i}\omega t}, \tag{5.16}$$

where $k_0 = \omega_0^2/g$ for deep water and $\omega = \omega_0 + k_0 U\cos\theta$ is the relative frequency. We consider short waves with respect to the ship length L, i.e., $k_0 L = \omega_0^2 L/g \gg 1$. However it is more convenient to choose $k = \omega^2/g$ as large parameter.

We introduce the well-known ray expansion

$$\phi(\mathbf{x},t;k)=a(\mathbf{x},k)\mathrm{e}^{\mathrm{i}kS(\mathbf{x})+\mathrm{i}\omega t}, \tag{5.17}$$

where $S(\mathbf{x})$ is the phase function and $a(\mathbf{x},k)$ the amplitude function. The latter is written as a regular series expansion with respect to inverse powers of ik,

$$a(\mathbf{x},k)=\sum_{j=0}^{N}\frac{a_j(\mathbf{x})}{(\mathrm{i}k)^j}+\mathrm{o}((\mathrm{i}k)^{-N}). \tag{5.18}$$

We restrict ourselves to the determination of $S(\mathbf{x})$ and $a_0(\mathbf{x})$.

Insertion of (5.17) into the Laplace equation (5.15) gives

$$-k^2a\nabla_3S\cdot\nabla_3S+\mathrm{i}k(2\nabla_3a\cdot\nabla_3S+a\Delta_3S)+\mathrm{O}(1)=0. \tag{5.19}$$

The subscript 3 is used to indicate the three-dimensional ∇ and Δ operator. If no subscript is used the operator acting on S or a_0 is two-dimensional in the horizontal plane. Comparing orders of magnitude in (5.19) leads to a set of equations for S and a_0 to be satisfied in the fluid region:

$$\left.\begin{aligned}\mathrm{O}(k^2):&\quad \nabla_3S\cdot\nabla_3S=0\\ \mathrm{O}(k^1):&\quad 2\nabla_3a_0\cdot\nabla_3S+a_0\Delta_3S=0\end{aligned}\right\}\quad\text{in the fluid.} \tag{5.20}$$

Next we insert (5.17) into the free-surface condition (5.14) and obtain

$$\begin{aligned}&-k^2\{(1-\mathbf{u}\cdot\nabla S)^2-\mathrm{i}S_y\}a-\mathrm{i}k\{2\mathbf{u}\cdot\nabla a-2(\mathbf{u}\cdot\nabla S)(\mathbf{u}\cdot\nabla a)\\ &\quad-\mathbf{u}\cdot\nabla(\mathbf{u}\cdot\nabla S)a+\mathrm{i}a_y\}+\mathrm{O}(1)=0.\end{aligned} \tag{5.21}$$

Comparing orders of magnitude in (5.21) yields

$$\left.\begin{aligned}\mathrm{O}(k^2):&\quad \mathrm{i}S_y=(1-\mathbf{u}\cdot\nabla S)^2\\ \mathrm{O}(k^1):&\quad a_{0y}=\mathrm{i}\{2\mathbf{u}\cdot\nabla a_0-2(\mathbf{u}\cdot\nabla S)(\mathbf{u}\cdot\nabla a_0)-\mathbf{u}\cdot\nabla(\mathbf{u}\cdot\nabla S)a_0\}\end{aligned}\right\}\quad\text{at } y=0. \tag{5.22}$$

The equations for the phase function at the free surface is obtained by elimination of S_z. Equations (5.20) and (5.22) yield the eikonal equation

$$(1-\mathbf{u}\cdot\nabla S)^4-\nabla S\cdot\nabla S=0, \tag{5.23}$$

and the transport equation

$$\{2\nabla S+4(1-\mathbf{u}\cdot\nabla S)^3\mathbf{u}\}\cdot\nabla a_0+a_0MS=0, \tag{5.24}$$

where $MS=\Delta_3S-2\mathbf{u}\cdot\nabla(\mathbf{u}\cdot\nabla S)(1-\mathbf{u}\nabla S)^2$.

In order to solve the eikonal equation (5.23) we introduce the notation $\mathbf{p}=(p,q):=(S_x,S_z)$, write (5.23) in the standard form $F(x,z,S,p,q)=0$ and apply the method of characteristics. The equations for the characteristics are the Charpit-Lagrange equations:

$$\begin{aligned}
\frac{\mathrm{d}x}{\mathrm{d}\sigma} &= F_p = -4(1-\mathbf{u}\cdot\mathbf{p})^3 u - 2p,\\
\frac{\mathrm{d}z}{\mathrm{d}\sigma} &= F_q = -4(1-\mathbf{u}\cdot\mathbf{p})^3 w - 2q,\\
\frac{\mathrm{d}p}{\mathrm{d}\sigma} &= -(F_x + pF_S) = 4(1-\mathbf{u}\cdot\mathbf{p})^3(\mathbf{u}_x\cdot\mathbf{p}),\\
\frac{\mathrm{d}q}{\mathrm{d}\sigma} &= -(F_z + qF_S) = 4(1-\mathbf{u}\cdot\mathbf{p})^3(\mathbf{u}_z\cdot\mathbf{p}).
\end{aligned} \tag{5.25}$$

The solutions of these equations (5.25) are called *rays* as in geometrical optics. The phase function is obtained by solving the equation

$$\frac{\mathrm{d}S}{\mathrm{d}\sigma} = pF_p + qF_q = -4(1-\mathbf{u}\cdot\mathbf{p})^3 + 2\mathbf{p}\cdot\mathbf{p}. \tag{5.26}$$

One must realise that the rays are not perpendicular to the wave fronts $S = \text{constant}$. The transport equation along the rays becomes

$$\frac{\mathrm{d}a_0}{\mathrm{d}\sigma} = a_0 MS. \tag{5.27}$$

This operator MS has the final form

$$\begin{aligned}
MS = {} & S_{xx}\left\{1 - 2|\nabla S|u^2 - \frac{S_x^2}{S_x^2+S_z^2}\right\} + S_{xz}\left\{-4|\nabla S|u^2 - 2\frac{S_xS_z}{S_x^2+S_z^2}\right\}\\
& + S_{zz}\left\{1 - 2|\nabla S|u^2 - \frac{S_x^2}{S_x^2+S_z^2}\right\} - 2|\nabla S|\nabla(\mathbf{u}\cdot\mathbf{u})\cdot\nabla S.
\end{aligned} \tag{5.28}$$

Before we can solve the characteristic equations together with the phase and amplitude function, the second derivatives in MS must be determined. To obtain $S_{xx} = p_x$, $S_{xz} = p_z = q_x$ and $S_{zz} = q_z$ one may use numerical differentiation. On the other hand, ordinary differential equations for those terms can be derived, as well,

$$\begin{aligned}
\frac{\mathrm{d}p_x}{\mathrm{d}\sigma} &= 12(1-\mathbf{u}\cdot\mathbf{p})^2(\mathbf{u}\cdot\mathbf{p})_x^2 - 2\mathbf{p}_x\cdot\mathbf{p}_x - 4(1-\mathbf{u}\cdot\mathbf{p})^3(\mathbf{u}_{xx}\cdot\mathbf{p} + 2\mathbf{u}_x\cdot\mathbf{p}_x),\\
\frac{\mathrm{d}p_z}{\mathrm{d}\sigma} &= 12(1-\mathbf{u}\cdot\mathbf{p})^2(\mathbf{u}\cdot\mathbf{p})_x(\mathbf{u}\cdot\mathbf{p})_z - 2\mathbf{p}_x\cdot\mathbf{p}_z\\
&\quad - 4(1-\mathbf{u}\cdot\mathbf{p})^3(\mathbf{u}_{xz}\cdot\mathbf{p} + \mathbf{u}_x\cdot\mathbf{p}_z + \mathbf{u}_z\cdot\mathbf{p}_x),\\
\frac{\mathrm{d}q_z}{\mathrm{d}\sigma} &= 12(1-\mathbf{u}\cdot\mathbf{p})^2(\mathbf{u}\cdot\mathbf{p})_z^2 - 2\mathbf{p}_z\cdot\mathbf{p}_z - 4(1-\mathbf{u}\cdot\mathbf{p})^3(\mathbf{u}_{zz}\cdot\mathbf{p} + 2\mathbf{u}_z\cdot\mathbf{p}_z).
\end{aligned} \tag{5.29}$$

The characteristic equations together with the equations along these characteristics can be solved. We give initial conditions for the incident field at a distance from the object where the ray pattern is not disturbed by the double body potential. These ordinary differential equations are solved by RK4. At the object we take care of the proper reflection laws generated by the Neumann boundary condition (no flux).

The mean resistance $\mathbf{F}_{aw}$ is defined as the time-averaged force acting on the hull, due to waves. The force in the x-direction is the added resistance. In general we have

$$\mathbf{F}_{aw} = -\overline{\int_{y=-\infty}^{\zeta} \int_{\mathrm{WL}} p\mathbf{n}\,\mathrm{d}l\,\mathrm{d}y}. \tag{5.30}$$

In the asymptotic case this leads to the expression

$$\begin{aligned}\mathbf{F}_{aw} = &-\frac{1}{4}\int_{\mathrm{WL}} \{(\nabla S^{(i)} \cdot \nabla S^{(i)})^{\frac{1}{4}} a_0^{(i)} + (\nabla S^{(r)} \cdot \nabla S^{(r)})^{\frac{1}{4}} a_0^{(r)}\}^2 \mathbf{n}\,\mathrm{d}l \\ &+ \frac{1}{4}\int_{\mathrm{WL}} \Bigg\{ a_0^{(i)2} |\nabla S^{(i)}| + a_0^{(r)2} |\nabla S^{(r)}| \\ &+ 2a_0^{(i)} a_0^{(r)} \frac{\nabla S^{(i)} \cdot \nabla S^{(r)} + |\nabla S^{(i)}||\nabla S^{(r)}|}{|\nabla S^{(i)}| + |\nabla S^{(r)}|} \Bigg\} \mathbf{n}\,\mathrm{d}l.\end{aligned} \tag{5.31}$$

The superscripts for the amplitude and the phase indicate incoming and reflected waves. In Figs. 5.1 and 5.2 we give results of the ray pattern for a circular cylinder in deep water for $\theta = 0°$ and $\tau = 0.25$ and $\tau = 0.5$. In Fig. 5.3 the values of the mean

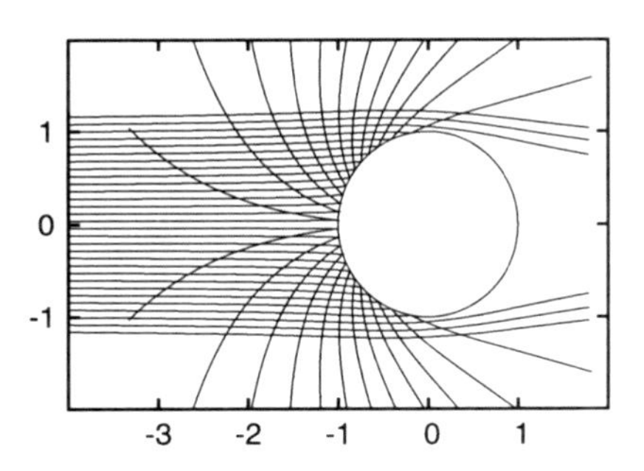

Fig. 5.1 Ray pattern for a cylinder with $\tau = 0.25$

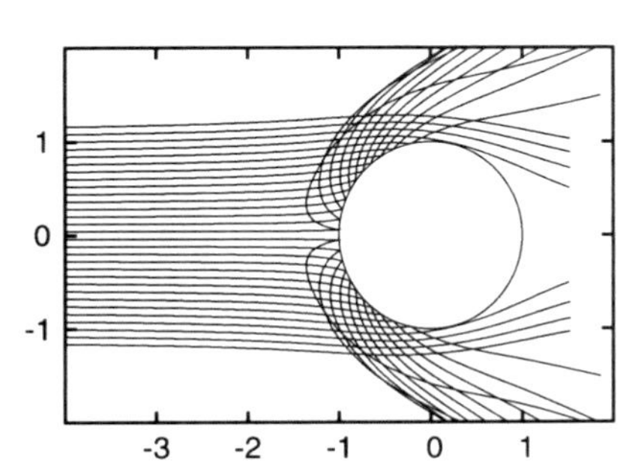

Fig. 5.2 Ray pattern for a cylinder with $\tau = 0.5$

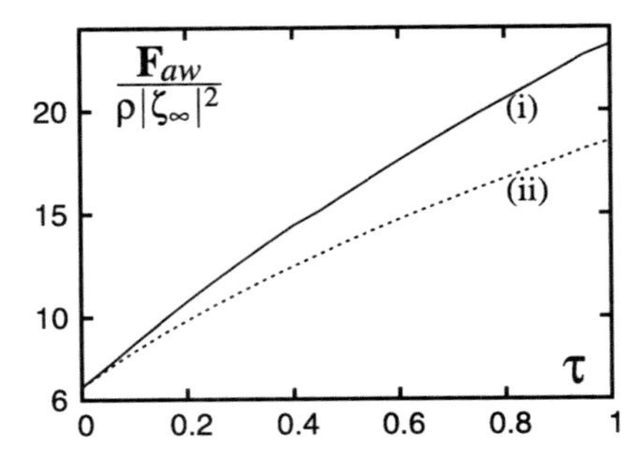

Fig. 5.3 Added resistance for (*i*) a circular cylinder and (*ii*) a sphere

forces are given for a circular cylinder and a sphere. In Fig. 5.2 we see that in front of the blunt bow the reflected rays form a caustic. The amplitude near this line becomes infinite. In principal one can derive a uniformly valid asymptotic theory with finite amplitude near this line. Computations show that the waves become shorter and the amplitude larger near the caustic. The result of this is that the wave breaks in front of the blunt bow, even in the case of low incident waves. This is observed in practice as well.

One may conclude that a proper description of the velocity field near the stagnation point influences the wave pattern near the bow greatly and that the added resistance increases significantly for increasing values of the velocity.

Chapter 6
Flexible Floating Platform

In this chapter we consider the two-dimensional interaction of an incident wave with a flexible floating dock or very large floating platform (VLFP) with finite draft. The water depth is finite. The case of a rigid dock is a classical problem. For instance Mei and Black [10] have solved the rigid problem, by means of a variational approach. They considered a fixed bottom and fixed free surface obstacle, so they also covered the case of small draft. After splitting the problem in a symmetric and an antisymmetric one, the method consists of matching of eigenfunction expansions of the velocity potential and its normal derivative at the boundaries of two regions. In principle, their method can be extended to the flexible platform case. Recently we derived a simpler method for both the moving rigid and the flexible dock [5]. However we considered objects with zero draft only. In this chapter we present our approach for the case of finite, but small, draft. The draft is small compared to the length of the platform to be sure that we may use as a model, for the elastic plate, the thin plate theory, while the water pressure at the plate is applied at finite depth. The method is based on a direct application of Green's theorem, combined with an appropriate choice of expansion functions for the potential in the fluid region outside the platform and the deflection of the plate. The integral equation obtained by the Green's theorem is transformed into an integral-differential equation by making use of the equation for the elastic plate deflection. One must be careful in choosing the appropriate Green's function. It is crucial to use a formulation of the Green's function consisting of an integral expression only. In Sect. 9.4 we derive such a Green's function for the two-dimensional case. One may derive an expression as can be found in the article of Wehausen and Laitone [19] after application of Cauchy's residue lemma. In the three-dimensional case one also may derive such an expression. The advantage of this version of the source function is that one may work out the integration with respect to the space coordinate first and apply the residue lemma afterwards. In the case of a zero draft platform this approach resulted in a dispersion relation in the plate region and an algebraic set of equations for the coefficients of the deflection only. Here we derive a coupled algebraic set of equations for the expansion coefficients of the potential in the fluid region and the deflection.

A.J. Hermans, *Water Waves and Ship Hydrodynamics*,
DOI 10.1007/978-94-007-0096-3_6, © Springer Science+Business Media B.V. 2011

6.1 The Finite Draft Problem

In this section we derive the general formulation for the diffraction of waves by a flexible platform of general geometric form. The fluid is ideal, so we introduce a velocity potential using $\mathbf{V}(\mathbf{x},t) = \nabla\Phi(\mathbf{x},t)$, where $\mathbf{V}(\mathbf{x},t)$ is the fluid velocity vector. Hence $\Phi(\mathbf{x},t)$is a solution of the Laplace equation

$$\Delta\Phi = 0 \quad \text{in the fluid,} \tag{6.1}$$

together with the linearised kinematic condition, $\Phi_y = \tilde{v}_t$, and dynamic condition, $p/\rho = -\Phi_t - g\tilde{v}$, at the mean water surface $y = 0$, where $\tilde{v}(x,z,t)$ denotes the free surface elevation, and ρ is the density of the water. The linearised free surface condition outside the platform, $y = 0$ and $(x,z) \in \mathscr{F}$, becomes

$$\frac{\partial^2\Phi}{\partial t^2} + g\frac{\partial\Phi}{\partial y} = 0. \tag{6.2}$$

The platform is situated at the mean free-surface $y = 0$, its thickness is d. The platform is modelled as an elastic plate with zero thickness. The neutral axis of the plate is at $y = 0$, while the water pressure distribution is applied at $y = -d$. Meylan et al. [11] have considered finite thickness as well. They consider the elastic equation for the deflection of a plate of finite thickness, however they apply the equation of motion at $y = 0$. They show for large platforms a minor influence due to the change of the elastic model. Our elastic model can easily be modified by changing the fourth-order differential operator, but due to lack of knowledge of suitable parameters we decided not to do so. So we neglect horizontal and torsional motion. To describe the vertical deflection $\tilde{v}(x,z,t)$, we apply the isotropic thin-plate theory, which leads to an equation for $\tilde{v}$ of the form

$$m(x,z)\frac{\partial^2\tilde{v}}{\partial t^2} = -\left(\frac{\partial^2}{\partial x^2} + \frac{\partial^2}{\partial z^2}\right)\left(D(x,z)\left(\frac{\partial^2\tilde{v}}{\partial x^2} + \frac{\partial^2\tilde{v}}{\partial z^2}\right)\right) + p|_{y=-d} \tag{6.3}$$

where $m(x,z)$ is the piece-wise constant mass of unit area of the platform while the piece-wise constant $D(x,z)$ is its equivalent flexural rigidity. We differentiate (6.3) with respect to t and use the kinematic and dynamic condition to arrive at the following equation for Φ at $y = -d$ in the platform area $(x,z) \in \mathscr{P}$:

$$\left\{\left(\frac{\partial^2}{\partial x^2} + \frac{\partial^2}{\partial z^2}\right)\left(\frac{D(x,z)}{\rho g}\left(\frac{\partial^2}{\partial x^2} + \frac{\partial^2}{\partial z^2}\right)\right) + \frac{m(x,z)}{\rho g}\frac{\partial^2}{\partial t^2} + 1\right\}\frac{\partial\Phi}{\partial y} + \frac{1}{g}\frac{\partial^2\Phi}{\partial t^2} = 0. \tag{6.4}$$

Due to the fact that the plate is freely floating we do not consider the hydrostatic pressure.

The edges of the platform are free of shear forces and moment. We assume that the flexural rigidity is constant along the edge and its derivative normal to the edge equals zero. Also, we assume that the radius of curvature, in the horizontal plane, of the edge is large. Hence, the edge may be considered to be straight locally. We then have the following boundary conditions at the edge:

$$\frac{\partial^2 \tilde{v}}{\partial n^2} + \nu \frac{\partial^2 \tilde{v}}{\partial s^2} = 0 \quad \text{and} \quad \frac{\partial^3 \tilde{v}}{\partial n^3} + (2-\nu)\frac{\partial^3 \tilde{v}}{\partial n \partial s^2} = 0 \tag{6.5}$$

where ν is Poisson's ratio, n is in the normal direction, in the horizontal plane, along the edge and s denotes the arc-length along the edge. At the bottom of the fluid region $y = -h$ we have

$$\frac{\partial \Phi}{\partial y} = 0. \tag{6.6}$$

We assume that the velocity potential is a time-harmonic wave function, $\Phi(\mathbf{x}, t) = \phi(\mathbf{x})e^{i\omega t}$. We introduce the following parameters:

$$K = \frac{\omega^2}{g}, \qquad \mu = \frac{m\omega^2}{\rho g}, \qquad \mathscr{D} = \frac{D}{\rho g}.$$

In a practical situation the total length L of the platform is a few thousand metres. We obtain at the free surface, $y = 0$,

$$\frac{\partial \phi}{\partial z} - K\phi = 0 \tag{6.7}$$

and at the plate, $y = -d$, for a single strip,

$$\left\{\mathscr{D}\left(\frac{\partial^2}{\partial x^2} + \frac{\partial^2}{\partial z^2}\right)^2 - \mu + 1\right\}\frac{\partial \phi}{\partial y} - K\phi = 0. \tag{6.8}$$

The potential of the undisturbed incident wave is given by

$$\phi^{inc}(\mathbf{x}) = \frac{g\eta_\infty}{i\omega}\frac{\cosh(k_0(y+h))}{\cosh(k_0 h)}\exp\{ik_0(x\cos\beta + z\sin\beta)\} \tag{6.9}$$

where η_∞ is the wave amplitude in the original coordinate system, ω the frequency, while the wave number k_0 is the negative real solution of the dispersion relation

$$k_0 \tanh(k_0 h) = K, \tag{6.10}$$

for finite water depth. We restrict ourselves to the case of normal incidence, $\beta = 0$. In [6] is shown that the extension to oblique waves can be done easily.

To obtain an integral equation for the deflection $\tilde{v}(x, z, t) = \Re[v(x, z)e^{i\omega t}]$ of the platform, see [4, 6], it is very convenient to apply the Green's theorem, making use of the Green's function, $\mathscr{G}(\mathbf{x}; \boldsymbol{\xi})$, that fulfils boundary conditions at the seabed (6.6) and at the free surface (6.7). Application of Green's theorem in the fluid domain leads to the following expression for the potential function:

$$4\pi\phi(\mathbf{x}) = 4\pi\phi^{inc}(\mathbf{x}) + \int_{\mathscr{C}} \phi(\boldsymbol{\xi})\frac{\partial\mathscr{G}(\mathbf{x},\boldsymbol{\xi})}{\partial n}\,\mathrm{d}S + \int_{\mathscr{P}}\left(\phi(\boldsymbol{\xi})\frac{\partial\mathscr{G}(\mathbf{x},\boldsymbol{\xi})}{\partial\eta} - \frac{\partial\phi(\boldsymbol{\xi})}{\partial\eta}\mathscr{G}(\mathbf{x};\boldsymbol{\xi})\right)\mathrm{d}S. \tag{6.11}$$

The first integral is along the vertical sides of the platform, where the normal velocity of the fluid equals zero. The second integral is along the flat bottom. In the two-dimensional case, (x, y)-plane, the expression for the total potential becomes

$$\begin{aligned} 2\pi\phi(x,y) &= 2\pi\phi^{inc}(x,y) \\ &+ \int_{-d}^{0}\left(\phi(0,\eta)\frac{\partial\mathscr{G}(x,y;0,\eta)}{\partial\xi} - \phi(l,\eta)\frac{\partial\mathscr{G}(x,y;l,\eta)}{\partial\xi}\right)\mathrm{d}\eta \\ &+ \int_{0}^{l}\left(\phi(\xi,-d)\frac{\partial\mathscr{G}(x,y;\xi,-d)}{\partial\eta} - \frac{\partial\phi(\xi,-d)}{\partial\eta}\mathscr{G}(x,y;\xi,-d)\right)\mathrm{d}\xi. \end{aligned} \tag{6.12}$$

We continue with the two-dimensional case.

The Green's function $G(x, y; \xi, \eta)$ for the two dimensional case can be derived by means of a Fourier transform with respect to the x-coordinate. As is shown in Sect. 9.4 it has the form:

$$G(x,y;\xi,\eta) = \int_{-\infty}^{\infty}\frac{1}{\gamma}\frac{K\sinh\gamma y + \gamma\cosh\gamma y}{K\cosh\gamma h - \gamma\sinh\gamma h}\cosh\gamma(\eta+h)\mathrm{e}^{\mathrm{i}\gamma(x-\xi)}\,\mathrm{d}\gamma \quad \text{for } y > \eta \tag{6.13}$$

and

$$G(x,z;\xi,\eta) = \int_{-\infty}^{\infty}\frac{1}{\gamma}\frac{K\sinh\gamma\eta + \gamma\cosh\gamma\eta}{K\cosh\gamma h - \gamma\sinh\gamma h}\cosh\gamma(y+h)\mathrm{e}^{\mathrm{i}\gamma(x-\xi)}\,\mathrm{d}\gamma \quad \text{for } y < \eta. \tag{6.14}$$

If we close the contour of integration in the complex γ-plane we obtain the complex version of formula (13.34), as can be found in Wehausen and Laitone [19]

$$G(x,z;\xi,\eta) = -2\pi\mathrm{i}\sum_{i=0}^{\infty}\frac{1}{k_i}\frac{k_i^2 - K^2}{hk_i^2 - hK^2 + K}\cosh k_i(y+h)\cosh k_i(\eta+h)\mathrm{e}^{\mathrm{i}k_i|x-\xi|}, \tag{6.15}$$

where k_0 and $k_i, i = 1, \ldots, \infty$ are the positive real and positive imaginary zeros of the dispersion relation (6.10).

The advantage of this formulation for the Green's function is that, by means of the Green's theorem, we can derive an algebraic set of equations for the expansion coefficients by carrying out the integration with respect to the spatial variable analytically first.

It is well known that for the rigid case, Mei and Black [10], the potential can be expanded in eigenfunctions in the regions outside and underneath the platform. In the traditional approach, continuity of mass and velocity leads to sets of equations at $x = 0$ and $x = l$ respectively. The use of orthogonality relations then gives a set of equations for the unknown coefficients. In the case of zero thickness it is shown by Hermans [5] that a set of algebraic equations can be obtained for the expansion coefficients of the deflection alone. Here we also use this approach to obtain a coupled set of algebraic equations for the finite thickness case as well. It is also possible to make a non-orthogonal expansion, see for instance [8], of the potential underneath the flexible platform. In that case one can express, a posteriori, the deflection as an expansion in exponential functions. The dispersion relations derived by both approaches are the same, as expected.

6.2 Semi-Analytic Solution

Equation (6.12) and or the three-dimensional version (6.11), together with the condition at the bottom of the plate (6.8), can be solved by means of a numerical diffraction code. However, it is interesting to see how one can solve the equations semi-analytically for simple geometries. Here we work out the case of a strip.

We eliminate in relation (6.12) the function $\phi(\xi, -d)$ by using (6.8) and the kinematic condition

$$\phi_\eta(\xi, -d) = -\mathrm{i}\omega v(\xi). \tag{6.16}$$

Thus we obtain

$$\begin{aligned} 2\pi\phi(x, y) = 2\pi\phi^{inc}(x, y) &+ \int_{-d}^{0} \left(\phi(0, \eta) \frac{\partial \mathscr{G}(x, y; 0, \eta)}{\partial \xi} \right. \\ &\left. - \phi(l, \eta) \frac{\partial \mathscr{G}(x, y; l, \eta)}{\partial \xi} \right) \mathrm{d}\eta \\ &- \mathrm{i}\omega \int_0^l \left(\frac{1}{K} \left(\mathscr{D} \frac{\partial^4}{\partial \xi^4} - \mu + 1 \right) v(\xi) \frac{\partial \mathscr{G}(x, y; \xi, -d)}{\partial \eta} \right. \\ &\left. - v(\xi) \mathscr{G}(x, y; \xi, -d) \right) \mathrm{d}\xi. \end{aligned} \tag{6.17}$$

We assume that the deflection $v(x)$ can be written as an expansion in exponential functions, truncated at $N + 2$ terms of the form

$$v(x) = \eta_\infty \sum_{n=0}^{N+1} \left(a_n \mathrm{e}^{\mathrm{i}\kappa_n x} + b_n \mathrm{e}^{-\mathrm{i}\kappa_n (x-l)} \right). \tag{6.18}$$

The values for κ_n follow from a 'dispersion' relation, yet to be determined. If we consider κ_n's with either real negative values or, if they are complex, with positive imaginary part, then the first part of expression (6.18) expresses modes travelling and evanescent to the right. The second part then describes modes travelling and evanescent to the left.

Furthermore we expand the potential function for $x \leq 0$ and $x \geq l$ in series of orthogonal eigen-functions, truncated at N terms

$$\phi(x,y) = \frac{g\eta_\infty}{\mathrm{i}\omega}\left(\frac{\cosh k_0(y+h)}{\cosh k_0 h}\mathrm{e}^{\mathrm{i}k_0 x} + \sum_{n=0}^{N-1}\alpha_n \frac{\cosh k_n(y+h)}{\cosh k_n h}\mathrm{e}^{-\mathrm{i}k_n x}\right) \quad \text{for } x \leq 0 \tag{6.19}$$

and

$$\phi(x,y) = \frac{g\eta_\infty}{\mathrm{i}\omega}\sum_{n=0}^{N-1}\beta_n \frac{\cosh k_n(y+h)}{\cosh k_n h}\mathrm{e}^{\mathrm{i}k_n(x-l)} \quad \text{for } x \geq l. \tag{6.20}$$

The difference in the number of expansion functions in (6.18) is due to the fact that we have four boundary conditions at the edge of the plate (6.5). The coefficients α_0 and β_0 are the reflection and transmission coefficients respectively. it should be noticed that the potential under the platform is **not** expanded in a set of orthogonal eigen-functions. By the way, such a set does not exist. Extension of the solution along the bottom of the platform in the flow region is simply done by application of (6.17). We have introduced $4N+4$ unknown coefficients. Next we derive an algebraic set of equations for these coefficients.

First we take (x, y) at the bottom of the plate, this leads to the following equation:

$$\begin{aligned} 2\pi\left(\mathscr{D}\frac{\partial^4}{\partial x^4} - \mu + 1\right)(x) &= -2\pi\frac{K}{\mathrm{i}\omega}\phi^{inc}(x,-d) \\ &\quad - \frac{K}{\mathrm{i}\omega}\int_{-d}^{0}\left(\phi(0,\eta)\frac{\partial\mathscr{G}(x,-d;0,\eta)}{\partial\xi} - \phi(l,\eta)\frac{\partial\mathscr{G}(x,-d;l,\eta)}{\partial\xi}\right)\mathrm{d}\eta \\ &\quad + \lim_{y\uparrow -d}\int_0^l\left(\left(\mathscr{D}\frac{\partial^4}{\partial\xi^4} - \mu + 1\right)v(\xi)\frac{\partial\mathscr{G}(x,y;\xi,-d)}{\partial\eta} \right. \\ &\quad \left. - Kv(\xi)\mathscr{G}(x,y;\xi,-d)\right)\mathrm{d}\xi. \end{aligned} \tag{6.21}$$

We take the limit in the last integral after we have carried out the spatial integrations analytically. This means that we keep the factor 2π in the left-hand side of the equation. The commonly used factor π and principle value integral may be obtained by taking the limit first. However, it is more convenient to avoid the principle value integral in our approach. In the first integral on the right-hand side we insert for the Green function the series expansion (6.15) and for the potential function the expansions (6.19) and (6.20), while in the second integral we use (6.14) for the Green function and (6.18) for the deflection. In the first integral integration with respect to η and in the last integral the integration with respect to ξ can be carried out. Next we close the remaining contour of integration in the complex γ-plane.

If we now equalise the coefficients of $\mathrm{e}^{\mathrm{i}\kappa_n x}$ and of $\mathrm{e}^{-\mathrm{i}\kappa_n (x-l)}$, we obtain the following 'dispersion' relation for κ_n, the κ_n's are the zero's of

$$(\mathscr{D}\kappa^4 - \mu)K \cosh\kappa d + (K^2 - \kappa^2(\mathscr{D}\kappa^4 - \mu + 1))\frac{\sinh\kappa d}{\kappa}$$
$$= (\mathscr{D}\kappa^4 - \mu + 1)\frac{(K\cosh\kappa h - \kappa\sinh\kappa h)}{\cosh\kappa(-d+h)}.$$

After some manipulations this relation can be rewritten in the form

$$((\mathscr{D}\kappa^4 - \mu - 1)\kappa\tanh\kappa(h-d) - K)(K\sinh\kappa d - \kappa\cosh\kappa d) = 0. \tag{6.22}$$

For $d = 0$ the dispersion relation for the zero draft platform is recovered. It should be noticed that relation (6.22) is not exactly the same as the zero draft relation with h replaced by $h - d$. Hence, we ignore the zeros of the second part, which occur for values of K sufficiently large only.

6.2.1 Semi-Infinite Platform

Let us first consider the half-plane problem. We introduce some slight physical damping to get rid of the contributions of the upper bound in the last integral in (6.21) and the second part of the first integral. The terms we obtain after closure of the contour in the last integral of (6.21) contain the exponential functions $\mathrm{e}^{\mathrm{i}k_n x}$. We take the coefficients of each exponential equal to zero. This leads to a set of N algebraic equations for the coefficients a_n and α_n. For the half-plane problem, we obtain for $i = 0, \ldots, N-1$:

$$\sum_{n=0}^{N-1}\frac{\alpha_n}{\cosh k_n h}\mathscr{K}_{i,n} - \sum_{n=0}^{N+1}\frac{a_n}{\kappa_n - k_i}\bigg((\mathscr{D}\kappa_n^4 - \mu + 1)\sinh k_i(h-d)$$
$$- \frac{K}{k_i}\cosh k_i(h-d)\bigg)$$
$$= \delta_i^0\frac{hk_0^2 - hK^2 + K}{(k_0^2 - K^2)\cosh k_0 h)} - \frac{\mathscr{K}_{i,0}}{\cosh k_0 h}, \tag{6.23}$$

where the coefficients $\mathscr{K}_{i,n}$ are defined as

$$2\mathscr{K}_{i,n} = \frac{1}{k_i + k_n}[\sinh(k_i + k_n)h - \sinh(k_i + k_n)(h-d))]$$
$$+ \frac{1}{k_i - k_n}[\sinh(k_i - k_n)h - \sinh(k_i - k_n)(h-d)]. \tag{6.24}$$

This is a set of N equations for $2N+2$ unknown coefficients. We have two conditions at the edge of the plate, so we must still obtain N. At the vertical front end of the platform (6.17) gives the relation

$$2\pi\phi(0,y) = 2\pi\phi^{inc}(0,y) + \lim_{x\to 0}\int_{-d}^{0}\phi(0,\eta)\frac{\partial\mathscr{G}(x,y;0,\eta)}{\partial\xi}\,d\eta$$
$$- i\omega\int_0^\infty\left(\frac{1}{K}\left(\mathscr{D}\frac{\partial^4}{\partial\xi^4} - \mu + 1\right)v(\xi)\frac{\partial\mathscr{G}(0,y;\xi,-d)}{\partial\eta}\right.$$
$$\left. - v(\xi)\mathscr{G}(0,y;\xi,-d)\right)d\xi. \tag{6.25}$$

We insert the series expansions (6.18) and (6.19) in this equation and compare the coefficients of $\cosh k_i(y+h)$. For the Green function we use expression (6.15) in both integrals.

We obtain for $i = 0, \ldots, N-1$,

$$\frac{hk_i^2 - hK^2 + K}{(k_i^2 - K^2)\cosh k_i h}\alpha_i - \sum_{n=0}^{N-1}\frac{\alpha_n}{\cosh k_n h}\mathscr{K}_{i,n}$$
$$- \sum_{n=0}^{N+1}\frac{a_n}{\kappa_n + k_i}\left((\mathscr{D}\kappa_n^4 - \mu + 1)\sinh k_i(h-d) - \frac{K}{k_i}\cosh k_i(h-d)\right)$$
$$= \frac{1}{\cosh k_0 h}\mathscr{K}_{i,0}. \tag{6.26}$$

We give some results for the absolute value of the amplitude of the deflection for a semi-infinite platform with a draft of two metres and water depth of ten metres. In Fig. 6.1 results are shown for three values of the deep water wave length, $\lambda = 2\pi/K = 150, 90, 30$ m respectively. As expected the amplitude increases with increasing values of the wave length. In Fig. 6.2 we show for $\lambda = 90$ m and water depth of ten metre the absolute value of the amplitude for several values of the draft, $d = 0, 2, 4, 6$ m. In Fig. 6.3 we show the influence of water depth on the amplitude of deflection. We have chosen $h = 100, 20, 10$ m, $d = 2$ m and a fixed frequency with $\lambda = 90$ m. The amplitude of the deflection increases for increasing water depth. To carry out computations for the larger values of water depth one must get rid of all hyperbolic sin and cosine functions in the formulation. This can be done by using standard formulas for these functions and by using the dispersion relation for the free surface water waves. By doing so one obtains very accurate results. In Fig. 6.4 we show the real part of the deflection for the same values of water depth, $d = 5$ m and fixed values of the wavelength $\lambda_0 = 2\pi/k_0 = 100$ m. We also have computed the absolute value of the amplitude of the wave elevation in front of the platform. The result is shown in Fig. 6.5. It is clearly shown that the elevation of the wave and the platform are discontinuous at $x = 0$. The amplitude of the reflected wave $\alpha_0 = 0.45657 - 0.43639i$.

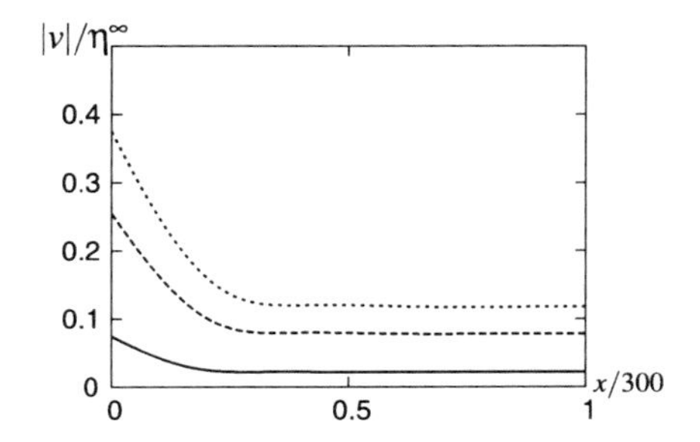

Fig. 6.1 $\mathscr{D} = 10^7$ m^4, $d = 2$ m, $h = 10$ m and $\lambda = 150, 90, 30$ m (top-down)

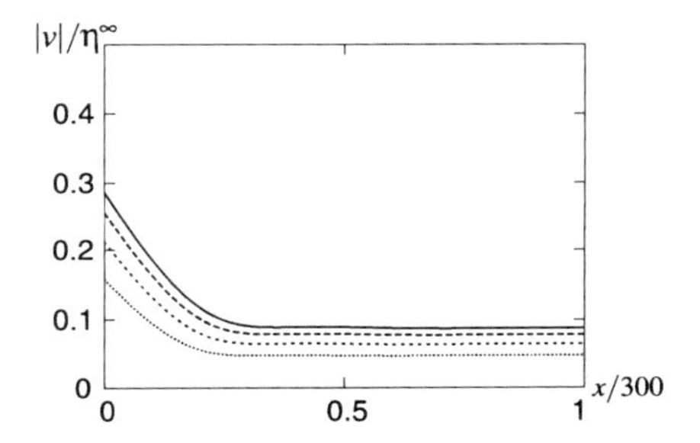

Fig. 6.2 $\mathscr{D} = 10^7$ m^4, $d = 0, 2, 4, 6$ (top-down), $h = 10$ m and $\lambda = 90$ m

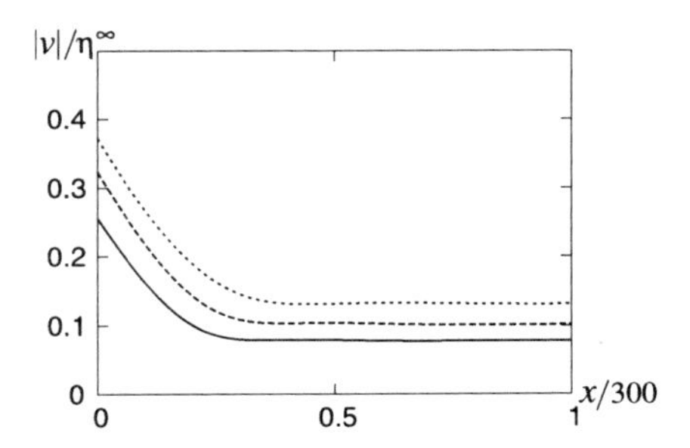

Fig. 6.3 $\mathscr{D} = 10^7$ m^4, $d = 2$ m, $h = 100, 20, 10$ m (top-down) and $\lambda = 90$ m

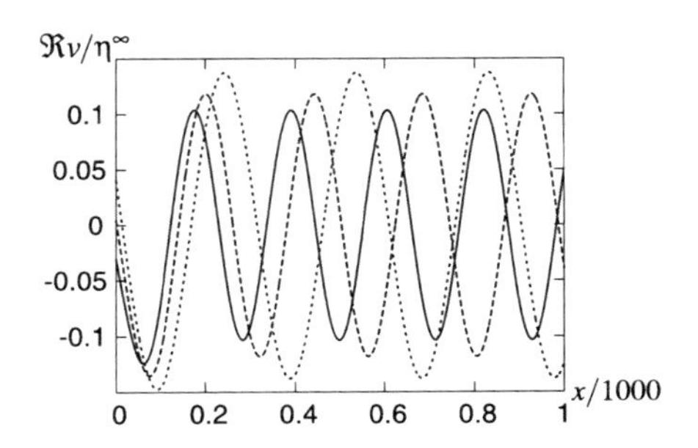

Fig. 6.4 Real part of the deflection for $\mathscr{D} = 10^7$ m^4, $d = 5$ m, $h = 100, 20, 10$ m (top-down) m and $\lambda_0 = 100$ m

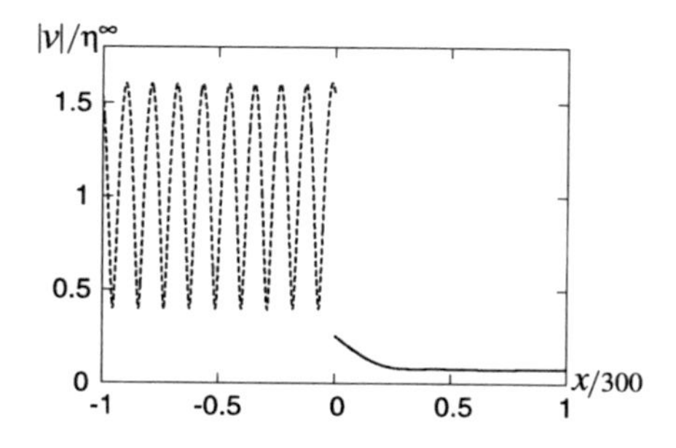

Fig. 6.5 Amplitude of wave and deflection for $\mathscr{D} = 10^7$ m^4, $d = 2$ m, $h = 10$ m and $\lambda = 90$ m

6.2.2 Strip of Finite Length

We follow the same procedure as for the semi-infinite case. The first step is to compare the coefficients of the exponential functions $\mathrm{e}^{\pm \mathrm{i}k_n x}$ in (6.21). This leads to a set of $2N$ algebraic equations for the coefficients a_n, b_n, α_n and β_n.

$$\begin{aligned}
&\sum_{n=0}^{N-1} \frac{\alpha_n}{\cosh k_n h} \mathscr{K}_{i,n} \\
&\quad - \sum_{n=0}^{N+1} \frac{a_n}{\kappa_n - k_i} \left((\mathscr{D}\kappa_n^4 - \mu + 1) \sinh k_i (h-d) - \frac{K}{k_i} \cosh k_i (h-d) \right) \\
&\quad + \sum_{n=0}^{N+1} \frac{b_n}{\kappa_n + k_i} \left((\mathscr{D}\kappa_n^4 - \mu + 1) \sinh k_i (h-d) - \frac{K}{k_i} \cosh k_i (h-d) \right) \mathrm{e}^{\mathrm{i}\kappa_n l} \\
&= \delta_i^0 \frac{hk_0^2 - hK^2 + K}{(k_0^2 - K^2)\cosh k_0 h)} - \frac{\mathscr{K}_{i,0}}{\cosh k_0 h}
\end{aligned} \tag{6.27}$$

and

$$\begin{aligned}
&\sum_{n=0}^{N-1} \frac{\beta_n}{\cosh k_n h} \mathscr{K}_{i,n} \\
&\quad + \sum_{n=0}^{N+1} \frac{a_n}{\kappa_n + k_i} \left((\mathscr{D}\kappa_n^4 - \mu + 1) \sinh k_i (h-d) - \frac{K}{k_i} \cosh k_i (h-d) \right) \mathrm{e}^{\mathrm{i}\kappa_n l} \\
&\quad - \sum_{n=0}^{N+1} \frac{b_n}{\kappa_n + k_i} \left((\mathscr{D}\kappa_n^4 - \mu + 1) \sinh k_i (h-d) - \frac{K}{k_i} \cosh k_i (h-d) \right) = 0.
\end{aligned} \tag{6.28}$$

This is a set of $2N$ equations for $4N + 4$ unknown coefficients. Next we consider the equations at $x = 0$ and $x = l$ respectively. After integration with respect to the spatial variable one obtains a summation of $\cosh k_i (y + h)$ terms. By taking the coefficients of each $\cosh k_i (y + h)$ function equal to zero we obtain the following set of $2N$ equations for the unknown expansion coefficients.

At $x = 0$ we get

$$
\begin{aligned}
&\frac{hk_i^2 - hK^2 + K}{(k_i^2 - K^2)\cosh k_i h}\alpha_i - \sum_{n=0}^{N-1} \frac{\alpha_n - \beta_n \mathrm{e}^{\mathrm{i}k_i l}}{\cosh k_n h}\mathscr{K}_{i,n} \\
&\quad - \sum_{n=0}^{N+1} \frac{a_n}{\kappa_n + k_i}\left((\mathscr{D}\kappa_n^4 - \mu + 1)\sinh k_i(h-d) - \frac{K}{k_i}\cosh k_i(h-d)\right) \\
&\times \left(1 - \mathrm{e}^{\mathrm{i}(\kappa_n + k_i)l}\right) \\
&\quad + \sum_{n=0}^{N+1} \frac{b_n}{\kappa_n - k_i}\left((\mathscr{D}\kappa_n^4 - \mu + 1)\sinh k_i(h-d) - \frac{K}{k_i}\cosh k_i(h-d)\right) \\
&\times \left(\mathrm{e}^{\mathrm{i}\kappa_n l} - \mathrm{e}^{\mathrm{i}k_i l}\right) \\
&= \frac{1}{\cosh k_0 h}\mathscr{K}_{i,0},
\end{aligned}
\tag{6.29}
$$

and at $x = l$ we get

$$
\begin{aligned}
&\frac{hk_i^2 - hK^2 + K}{(k_i^2 - K^2)\cosh k_i h}\beta_i + \sum_{n=0}^{N-1} \frac{\alpha_n \mathrm{e}^{\mathrm{i}k_i l} - \beta_n}{\cosh k_n h}\mathscr{K}_{i,n} \\
&\quad + \sum_{n=0}^{N+1} \frac{a_n}{\kappa_n - k_i}\left((\mathscr{D}\kappa_n^4 - \mu + 1)\sinh k_i(h-d) - \frac{K}{k_i}\cosh k_i(h-d)\right) \\
&\times \left(\mathrm{e}^{\mathrm{i}\kappa_n l} - \mathrm{e}^{\mathrm{i}k_i l}\right) \\
&\quad - \sum_{n=0}^{N+1} \frac{b_n}{\kappa_n + k_i}\left((\mathscr{D}\kappa_n^4 - \mu + 1)\sinh k_i(h-d) - \frac{K}{k_i}\cosh k_i(h-d)\right) \\
&\times \left(1 - \mathrm{e}^{\mathrm{i}(\kappa_n + k_i)l}\right) \\
&= \frac{hk_0^2 - hK^2 + K}{(k_0^2 - K^2)\cosh k_0 h}\mathrm{e}^{\mathrm{i}k_0 l} - \frac{1}{\cosh k_0 h}\mathscr{K}_{i,0}.
\end{aligned}
\tag{6.30}
$$

Together with the four relations at the end of the strip we have $4N + 4$ linear algebraic equations for the $4N + 4$ unknown coefficients.

The set of equations as is written here is not very suitable for numerical computations directly. Especially for large values of water-depth the arguments of the hyperbolic sine and cosine functions become rather large. So one is subtracting very large values in the computation of the coefficients. To obtain high numerical accuracy one must get rid of these functions. This can be done by using the dispersion relation for the water region. In Sect. 9.5 a more suitable set of equations is given.

We show some computational results for a two-dimensional platform of width 300 m. In all cases we take a fixed value for the flexural rigidity $\mathscr{D} = 10^7$ m^4, the

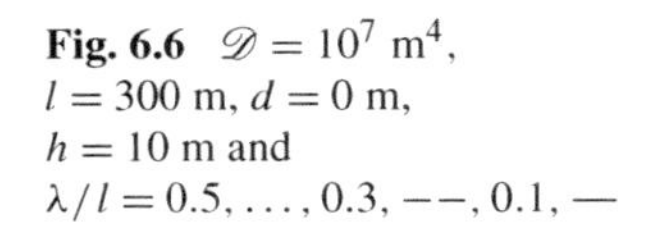

Fig. 6.6 $\mathscr{D} = 10^7$ m^4, $l = 300$ m, $d = 0$ m, $h = 10$ m and $\lambda/l = 0.5, \ldots, 0.3, --, 0.1,$ —

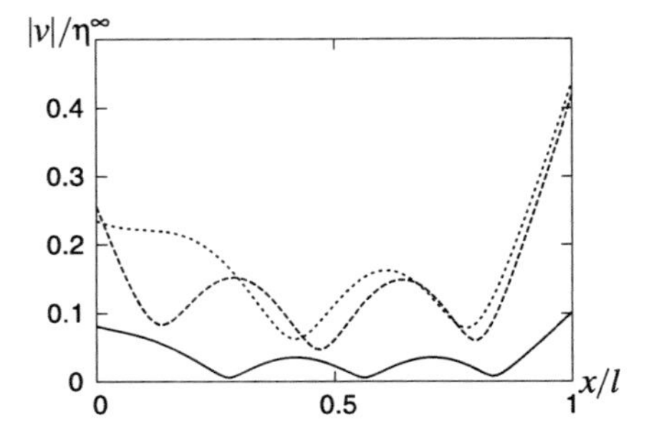

Fig. 6.7 $\mathscr{D} = 10^7$ m^4, $l = 300$ m, $d = 2$ m, $h = 10$ m and $\lambda/l = 0.5, 0.3, 0.1$

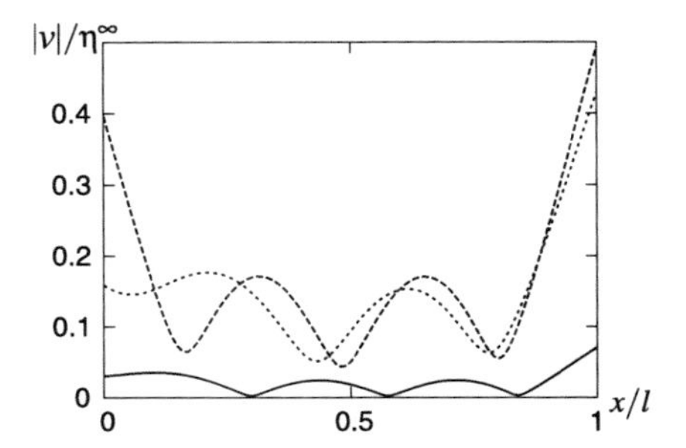

width of the strip $l = 300$ m and the water depth $h = 10$ m. In Figs. 6.6 and 6.7 we show for $d = 0$ and for $d = 2$ m the variation of the amplitude of deflection with respect to the wave length. In Figs. 6.8 and 6.9 the dependence on the draft for fixed values of the wave length is shown. The results of the first case show an increase of the deflection for increasing values of the draft. It will be shown later that this is due to a shift in the reflection curve. In Fig. 6.10 a result is shown for a larger value of the flexural rigidity $\mathscr{D} = 10^{10}$ m^4 and wave length $\lambda/l = 0.5$. This case is comparable with the interaction of free-surface waves with a rigid body. One clearly observes that the motion of the dock consists of a heave and pitch motion only.

In Figs. 6.11 and 6.13 we show for two values of the wave length the absolute value of the amplitude of the water surface in front of and behind the strip, together with the amplitude of the plate deflection for the zero draft case. The second case is near the zero reflection situation. In the four metre draft case, see Figs. 6.12 and 6.14, we see that $\lambda/l = 0.215$, or in terms of the actual wave length $\lambda_0/l = 0.178$, is close to total reflection. This is in contrast with the zero draft case in Fig. 6.13, due to the shift in the transmission-reflection curves. For the same reason the absolute value of the deflection increases if the draft increases in Fig. 6.8 in contrast with the result in Fig. 6.9.

The reflection and transmission coefficients for a strip of 300 m and depth 10 m are shown in Fig. 6.15, for zero draft and in Fig. 6.16 for a draft of 2 meteors. If we define $R = \alpha_0$ and $T = \beta_0$, notice no exponential function, we find that in all cases the relations $|T|^2 + |R|^2 = 1$ and $T\overline{R} + \overline{T}R = 0$, see for instance Mei et al. [10] or for a derivation Roseau [16], are fulfilled for at least 10 decimals. The coefficients are presented as a function of the actual wave length, $\lambda_0/l = 2\pi/k_0 l$. In Figs. 6.17

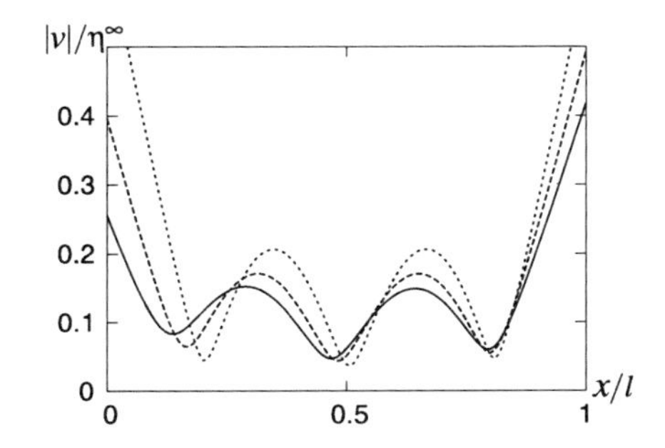

Fig. 6.8 $\mathscr{D} = 10^7$ m^4, $l = 300$ m, $d = 0$, —, 2, – –, 4, ... m, $h = 10$ m and $\lambda/l = 0.3$

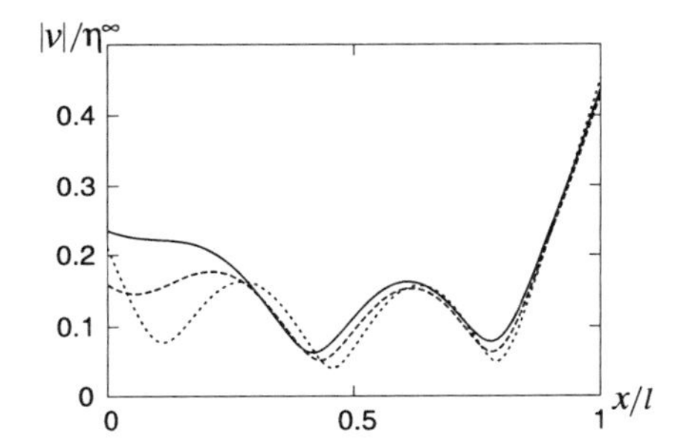

Fig. 6.9 $\mathscr{D} = 10^7$ m^4, $l = 300$ m, $d = 0$, —, 2, – –, 4, ... m, $h = 10$ m and $\lambda/l = 0.5$

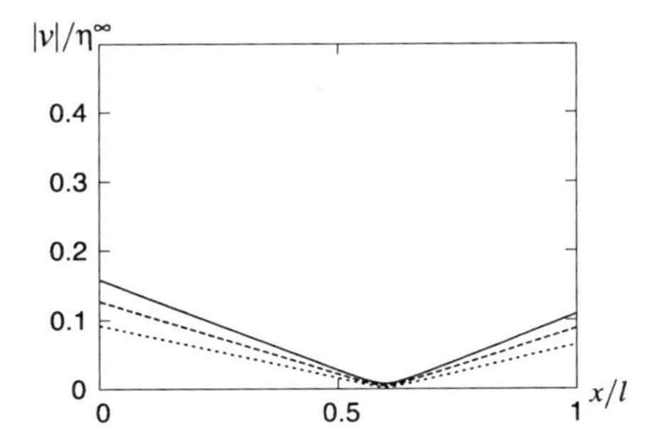

Fig. 6.10 $\mathscr{D} = 10^{10}$ m^4, $l = 300$ m, $d = 0, 2, 4$ m, $h = 10$ m and $\lambda/l = 0.5$

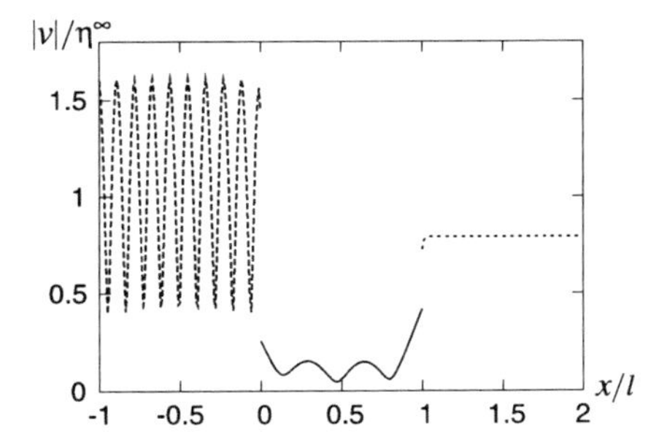

Fig. 6.11 $\mathscr{D} = 10^7$ m^4, $l = 300$ m, $d = 0$ m, $h = 10$ m and $\lambda/l = 0.3$

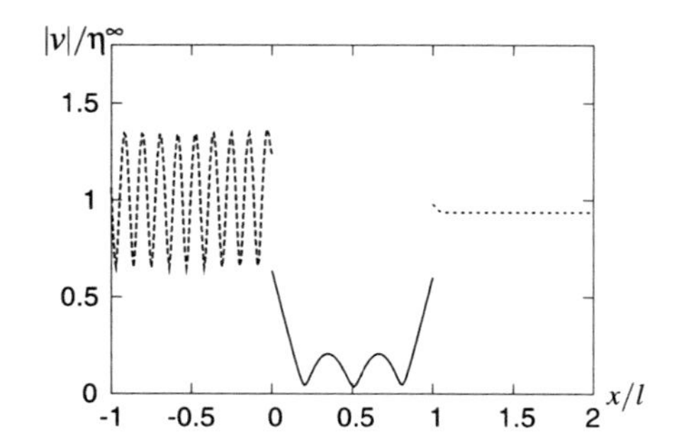

Fig. 6.12 $\mathscr{D} = 10^7$ m^4, $l = 300$ m, $d = 4$ m, $h = 10$ m and $\lambda/l = 0.3$

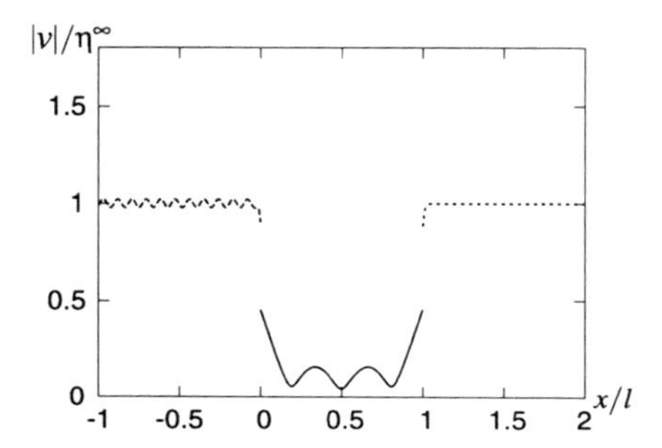

Fig. 6.13 $\mathscr{D} = 10^7$ m^4, $l = 300$ m, $d = 0$ m, $h = 10$ m and $\lambda/l = 0.215$

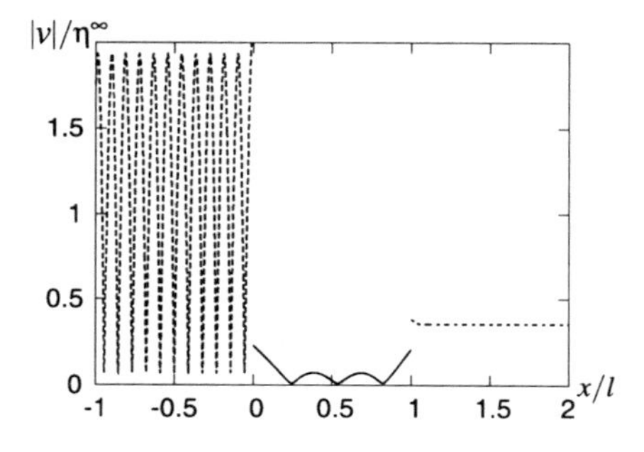

Fig. 6.14 $\mathscr{D} = 10^7$ m^4, $l = 300$ m, $d = 4$ m, $h = 10$ m and $\lambda/l = 0.215$

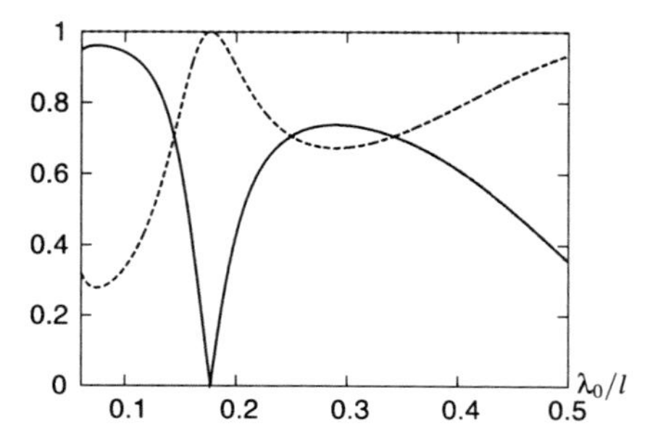

Fig. 6.15 —— Reflection and – – transmission coefficients for $h = 10$ m, $d = 0$ m and $l = 300$ m

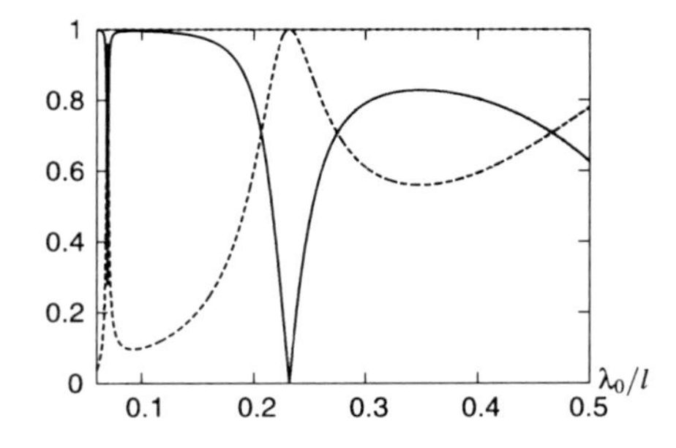

Fig. 6.16 Reflection and transmission coefficients for $h = 10$ m, $d = 4$ m and $l = 300$ m

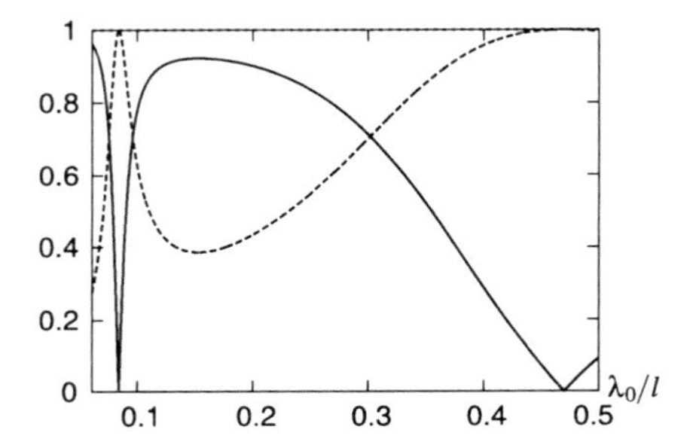

Fig. 6.17 Reflection and transmission coefficients for $h = 100$ m, $d = 0$ m and $l = 300$ m

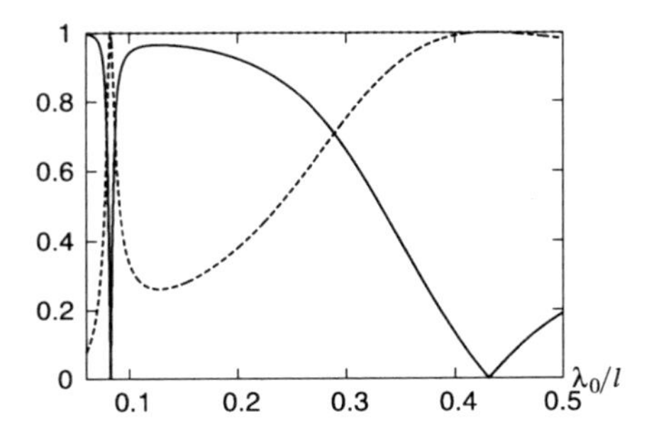

Fig. 6.18 Reflection and transmission coefficients for $h = 100$ m, $d = 2$ m, and $l = 300$ m

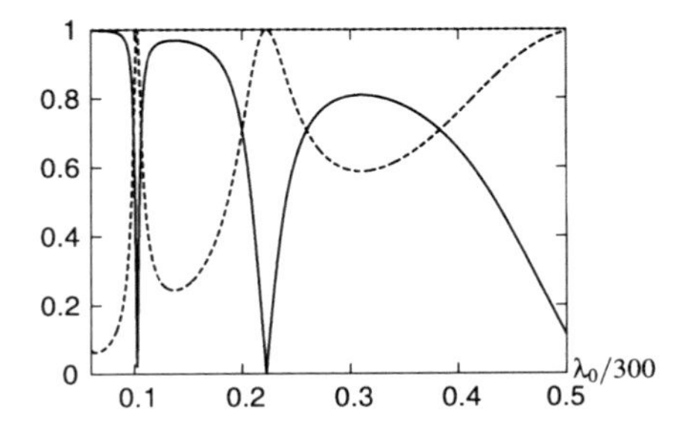

Fig. 6.19 Reflection and transmission coefficients for $h = 100$ m, $d = 2$ m, and $l = 650$ m

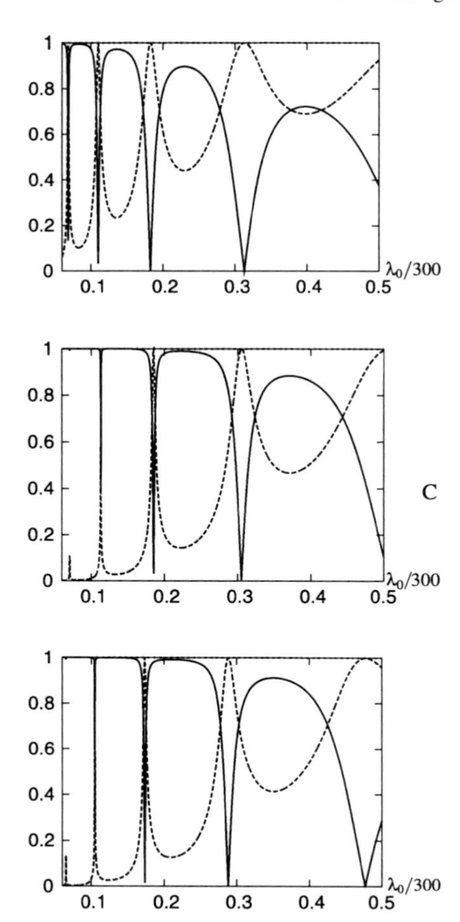

Fig. 6.20 Reflection and transmission coefficients for $h = 100$ m, $d = 2$ m and $l = 1000$ m

Fig. 6.21 Reflection and transmission coefficients for $h = 100$ m, $d = 8$ m, and $l = 1000$ m

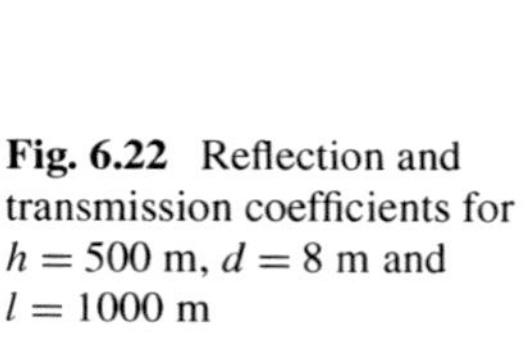

Fig. 6.22 Reflection and transmission coefficients for $h = 500$ m, $d = 8$ m and $l = 1000$ m

and 6.18 these coefficients are given for a water depth of 100 metres. In all cases the coefficient of flexural rigidity equals $\mathscr{D} = 10^7$ m^4. Figures 6.18, 6.19 and 6.20 show the results for different sizes of the strip. In Figs. 6.21 and 6.22 the result is shown for a strip of width $l = 100$ m and draft $d = 8$ m. It is clearly observed that for the short waves total reflection takes place.

Chapter 7
Irregular and Non-linear Waves

The surface waves of the sea are almost always random in the sense that detailed configuration of the surface varies in an irregular manner in both space and time. Section 7.1 contains a brief description of the Wiener spectrum in connection with the generalised Fourier representations for the surface waves [2, 20]. In this way we see how one may represent the surface elevation by a superposition of harmonic waves with amplitudes being a stochastic process.

The remaining sections in the chapter are devoted to non-linear waves. In Sect. 7.2 we give a systematic derivation of the shallow water theory from the exact hydrodynamical equations as the approximation of lowest order in a perturbation procedure. Here the relevant small parameter is the ratio of the depth of water to some characteristic length associated with the horizontal direction such as the wave length; the water is considered shallow when this parameter is small. It is a different kind of approximation from the previous linear theory for waves of small amplitude. The resulting equations here are quasi-linear and are exactly analogous to the ones in gas dynamics. Second order approximations are included in the last Sect. 7.3. In particular, an asymptotic theory will be developed for slowly varying wave trains, which may be considered as nearly uniform in the regions of order of magnitude of a small number of wave lengths and periods. Some non-linear dispersive wave phenomena will be discussed and more details can be found in [7].

7.1 Wiener Spectrum

The actual motion of the sea is by no means a harmonic motion with constant frequency. In Sect. 2.1.2, we gave a representation of the sea surface $\eta(x, t)$ in the form of an integral*

$$\eta(x, t) = \int_{-\infty}^{\infty} \left\{ A(k)\mathrm{e}^{-\mathrm{i}(kx-\omega t)} + A^*(k)\mathrm{e}^{\mathrm{i}(kx-\omega t)} \right\} \mathrm{d}k$$

under the hypothesis that $\int_{-}^{\infty} \infty |A(k)|^2 \, \mathrm{d}k$ is convergent. It is, however, not appropriate to suppose that the amplitudes of the high frequencies go to zero. The

A.J. Hermans, *Water Waves and Ship Hydrodynamics*,
DOI 10.1007/978-94-007-0096-3_7,

representation by a Fourier integral there breaks down and we must use a theory developed by Wiener of which we shall give an outline below. Details can be found in Wiener's book [20] in the references.

For a better understanding, we begin with a trigonometric polynomial $f(t)$ defined by

$$f(t) = \sum_{j=1}^{n} A_j e^{i\lambda_j t},$$

and calculate

$$f(t+\tau)f^*(t) = \sum_{j=1}^{n}\sum_{k=1}^{n} A_j A_k^* e^{i\lambda_j t} e^{i(\lambda_j - \lambda_k)\tau}.$$

Here A_k^* again denotes the complex conjugate of A_k. Now by taking the averaged mean of this quantity, we obtain that

$$\lim_{T\to\infty} \frac{1}{2T}\int_{-T}^{T} f(t+\tau)f^*(\tau)\,d\tau = \sum_{j=1}^{n}\sum_{k=1}^{n} A_j A_k^* e^{i\lambda_j t} \lim_{T\to\infty}\frac{1}{2T}\int_{-T}^{T} e^{i(\lambda_j-\lambda_k)\tau}\,d\tau.$$

For $\lambda_j = \lambda_k$, the limit is equal to unity, but for $\lambda_j \neq \lambda_k$, the limit vanishes. This leads to the result

$$\lim_{T\to\infty} \frac{1}{2T}\int_{-T}^{T} f(t+\tau)f^*(\tau)\,d\tau = \sum_{j=1}^{n} |A_j|^2 e^{i\lambda t}.$$

In other words, if we put

$$\varphi(t) = \lim_{T\to\infty} \frac{1}{2T}\int_{-T}^{T} f(t+\tau)f^*(\tau)\,d\tau,$$

then $\varphi(t)$ exists for every t, is continuous and consists of terms with the same frequency as those constituting $f(t)$ and with amplitudes equal to the square of the amplitudes of the corresponding terms of $f(t)$. Hence we see a way of finding a Fourier integral representation of a function which does not vanish at infinity. One can analyse the surface waves by means of functions of this kind in space variables as well as in the time variable.

We next build up the Wiener theory for those complex functions $f(t)$ of he real variable t such that

$$\varphi(t) = \lim_{T\to\infty} \frac{1}{2T}\int_{-T}^{T} f(t+\tau)f^*(\tau)\,d\tau, \tag{7.1}$$

exists for all t. Clearly then $\varphi(0)$ also exists. This means that the quadratic mean of the function exists. Moreover from the Schwartz's inequality, it can be shown that for all real t,

$$|\varphi(t)| \leq \varphi(0).$$

Indeed, it is easy to see that

$$
\begin{aligned}
|\varphi(t)| &= \overline{\lim_{T\to\infty}}\left|\frac{1}{2T}\int_{-T}^{T} f(t+\tau)f^*(\tau)\,\mathrm{d}\tau\right| \\
&\le \left\{\overline{\lim_{T\to\infty}}\frac{1}{2T}\int_{-T-|t|}^{T+|t|}|f(\tau)|^2\,\mathrm{d}\tau\,\overline{\lim_{T\to\infty}}\frac{1}{2T}\int_{-T}^{T}|f(\tau)|^2\,\mathrm{d}\tau\right\}^{1/2} \\
&\le \left\{\left(\lim_{T\to\infty}\frac{T+|t|}{T}\right)\left(\overline{\lim_{T\to\infty}}\frac{1}{2(T+|t|)}\int_{-T-|t|}^{T+|t|}|f(\tau)|^2\,\mathrm{d}\tau\right)\right. \\
&\qquad \left.\cdot\left(\overline{\lim_{T\to\infty}}\frac{1}{2T}\int_{-T}^{T}|f(\tau)|^2\,\mathrm{d}\tau\right)\right\}^{1/2} \\
&= \left\{1\times\varphi(0)\times\varphi(0)\right\}^{1/2} = \varphi(0).
\end{aligned}
$$

Here we use the notation $\overline{\lim} = \limsup$ for the *limit superior*. Recall that $\overline{\lim} S_n = A \iff$ the sequence $\{S_n\}_{n=1}^{\infty}$ is bounded above and has A as its largest limit point.

We now try to define a reasonable Fourier transform for those functions $f(t)$ in (7.1). Clearly from (7.1), the standard Fourier transform for $f(t)$, $\frac{1}{\sqrt{2\pi}}\int_{-\infty}^{\infty} f(\tau)\mathrm{e}^{-\mathrm{i}\omega\tau}\,\mathrm{d}\tau$, need not exist but the limit

$$\lim_{T\to\infty}\int_{-T}^{T} f(\tau)\mathrm{e}^{\mathrm{i}\omega\tau}\,\mathrm{d}\tau$$

does exist. This leads us to consider the integral

$$\int_{-\infty}^{\infty}\frac{\mathrm{e}^{\mathrm{i}\omega t}}{\mathrm{i}t}\,\mathrm{d}t,$$

although the integrand is singular at $t=0$. Since $\varphi(0)$ exists it follows that

$$\frac{1}{2T}\int_{-T}^{T}|f(t)|^2\,\mathrm{d}t$$

is bounded in T and hence the function which is $f(t)/\mathrm{i}t$ for $|t|>1$ and 0 for $t\le 1$ belongs to $\mathbf{L}_2$, consequently its Fourier transform (in $\mathbf{L}_2$) exists and belongs to $\mathbf{L}_2$. Thus we define

$$S(\omega) = \frac{1}{\sqrt{2\pi}}\,\underset{A\to\infty}{\mathrm{l.i.m.}}\left[\int_{-A}^{-1}+\int_{1}^{A}\right]\frac{f(t)\mathrm{e}^{-\mathrm{i}\omega t}}{-\mathrm{i}t}\,\mathrm{d}t + \frac{1}{\sqrt{2\pi}}\int_{-1}^{1}\frac{\mathrm{e}^{-\mathrm{i}\omega t}-1}{-\mathrm{i}t}\,\mathrm{d}t, \qquad (7.2)$$

and consider the difference $S(\omega+\varepsilon)-S(\omega-\varepsilon)$ which can be reduced to

$$S(\omega+\varepsilon)-S(\omega-\varepsilon) = \frac{1}{\sqrt{2\pi}}\,\underset{A\to\infty}{\mathrm{l.i.m.}}\int_{-A}^{A} f(t)\frac{2\sin\varepsilon t}{t}\mathrm{e}^{-\mathrm{i}\omega t}\,\mathrm{d}t. \qquad (7.3)$$

Here and in the sequel we use the notation for the limit in the mean: $\text{l.i.m.}_{\alpha\to\alpha_0} g_\alpha(\omega) = g(\omega)$ if

$$\lim_{\alpha\to\alpha_0}\int_{-\infty}^{\infty}|g_\alpha(\omega)-g(\omega)|^2\,\mathrm{d}\omega = 0,$$

where α_0 is any fixed constant including ∞. Because of our assumption for $f(t)$ in (7.1), clearly $S(\omega)$ in (7.2) exists and so is due to the difference $S(\omega+\varepsilon)-S(\omega-\varepsilon)$ in (7.3), which is the ordinary Fourier transform of $f(t)\frac{2\sin\varepsilon t}{t}$. Here $S(\omega)$ as defined by (7.2) is termed as the *generalised Fourier transform* of $f(t)$.

At this stage, Wiener uses a Tauberian therem: If $\varphi(t)\ge 0$ for $0\le t<\infty$, and either of the limits

$$\lim_{T\to\infty}\frac{1}{T}\int_0^T\varphi(t)\,\mathrm{d}t \quad\text{or}\quad \lim_{\varepsilon\to 0}\frac{2}{\pi\varepsilon}\int_0^\infty\varphi(t)\frac{\sin^2(\varepsilon t)}{t^2}\,\mathrm{d}t$$

exists, then the other limit exists and assumes the same value. Wiener's proof of this theorem is very complicated. However as a corollary, we have the result that

$$\lim_{\varepsilon\to 0}\frac{1}{4\pi\varepsilon}\int_{-\infty}^{\infty}|S(\omega+\varepsilon)-S(\omega-\varepsilon)|^2\,\mathrm{d}\omega = \lim_{T\to\infty}\frac{1}{2T}\int_{-T}^{T}|f(t)|^2\,\mathrm{d}t.$$

We now define $S_\tau(\omega)$, which bears the same relation to $f(t+\tau)$ as $S(\omega)$ to $f(t)$ in (7.2). Then we have

$$\begin{aligned}
&S_\tau(\omega+\varepsilon)-S_\tau(\omega-\varepsilon)-\mathrm{e}^{\mathrm{i}\tau\omega}\{S(\omega+\varepsilon)-S(\omega-\varepsilon)\}\\
&=\frac{1}{\sqrt{2\pi}}\left\{\underset{A\to\infty}{\text{l.i.m.}}\int_{-A}^{A}f(t+\tau)\frac{2\sin(\varepsilon t)}{t}\mathrm{e}^{-\mathrm{i}\varepsilon t}\,\mathrm{d}t\right.\\
&\left.\quad-\underset{A\to\infty}{\text{l.i.m.}}\int_{-A}^{A}f(t+\tau)\frac{2\sin(\varepsilon t)}{t}\mathrm{e}^{-\mathrm{i}\varepsilon(t-\tau)}\,\mathrm{d}t\right\}.
\end{aligned}$$

This can be transformed into

$$\frac{1}{\sqrt{2\pi}}\underset{A\to\infty}{\text{l.i.m.}}\int_{-A}^{A}f(t)\left[\frac{2\sin(\varepsilon(t-\tau))}{t-\tau}-\frac{2\sin(\varepsilon t)}{t}\right]\mathrm{e}^{-\mathrm{i}\omega(t-\tau)}\,\mathrm{d}t.$$

Thus from the Plancherel theorem, it follows that

$$\begin{aligned}
&\int_{-\infty}^{\infty}\left|S_\tau(\omega+\varepsilon)-S_\tau(\omega-\varepsilon)-\mathrm{e}^{\mathrm{i}\omega\tau}\{S(\omega+\varepsilon)-S(\omega-\varepsilon)\}\right|^2\mathrm{d}\omega\\
&=\int_{-\infty}^{\infty}|f(t)|^2\left[\frac{2\sin((t-\tau)\varepsilon)}{t-\tau}-\frac{2\sin(t\varepsilon)}{t}\right]^2\mathrm{d}t.
\end{aligned}$$

Now it can be shown that

$$\left|\frac{2\sin((t-\tau)\varepsilon)}{t-\tau}-\frac{2\sin(t\varepsilon)}{t}\right|\le\frac{16\varepsilon|\tau|}{|t|+|\tau|},$$

and hence by the Tauberian theorem and the fact that if $\frac{1}{2T}\int_{-t}^{T}|f(t)|^2\,\mathrm{d}t$ is bounded in T, then $\int_{-\infty}^{\infty}|f(t)|^2/(1+t^2)\,\mathrm{d}t<\infty$ and we have

$$\int_{-\infty}^{\infty}\left|S_\tau(\omega+\varepsilon)-S_\tau(\omega-\varepsilon)-\mathrm{e}^{\mathrm{i}\omega\tau}\left\{S(\omega+\varepsilon)-S(\omega-\varepsilon)\right\}\right|^2\mathrm{d}\omega=\mathrm{O}(\varepsilon^2).$$

Then from this, Wiener derives that

$$\varphi(\tau)=\lim_{\varepsilon\to 0}\frac{1}{4\pi\varepsilon}\int_{-\infty}^{\infty}\mathrm{e}^{\mathrm{i}\omega\tau}\left|S(\omega+\varepsilon)-S(\omega-\varepsilon)\right|^2\mathrm{d}\omega. \tag{7.4}$$

Putting

$$\varphi_\varepsilon(\tau)=\frac{1}{4\pi\varepsilon}\int_{-\infty}^{\infty}\mathrm{e}^{\mathrm{i}\omega\tau}\left|S(\omega+\varepsilon)-S(\omega-\varepsilon)\right|^2\mathrm{d}\omega,$$

Wiener shows that the function

$$\sigma_\varepsilon(\omega):=\frac{1}{\sqrt{2\pi}}\left\{\underset{A\to\infty}{\mathrm{l.i.m.}}\left[\int_1^A+\int_{-A}^{-1}\right]\frac{\varphi_\varepsilon(t)\mathrm{e}^{-\mathrm{i}\omega t}}{-\mathrm{i}t}\,\mathrm{d}t+\int_{-1}^{1}\varphi_\varepsilon(t)\frac{\mathrm{e}^{-\mathrm{i}\omega t}-1}{-\mathrm{i}t}\,\mathrm{d}t\right\}$$

is equal to

$$\text{constant}+\frac{1}{2\varepsilon\sqrt{2\pi}}\int_0^{\infty}|S(\omega'+\varepsilon)-S(\omega'-\varepsilon)|^2\,\mathrm{d}\omega'.$$

Or, with $\sigma(\omega)=\mathrm{l.i.m.}_{\varepsilon\to 0}\sigma_\varepsilon(\omega)$, we have

$$\sigma(\omega)-\sigma(-\omega)=\frac{1}{2\varepsilon\sqrt{2\pi}}\underset{\varepsilon\to 0}{\mathrm{l.i.m.}}\int_{-\omega}^{\omega}|S(\omega'+\varepsilon)+-S(\omega'-\varepsilon)|^2\,\mathrm{d}\omega'. \tag{7.5}$$

Finally, Wiener shows that

$$\varphi(\tau)=\frac{1}{\sqrt{2\pi}}\int_{-\infty}^{\infty}\mathrm{e}^{\mathrm{i}\tau\omega}\,\mathrm{d}\sigma(\omega). \tag{7.6}$$

If we let $F(\omega)\frac{1}{\sqrt{2\pi}}(\sigma(\omega)-\sigma(-\omega))$, then (7.6) can be written as

$$\varphi(\tau)=\int_0^{\infty}\cos(\omega\tau)\,\mathrm{d}F(\omega). \tag{7.7}$$

In this way we have introduced the spectral function $F(\omega)$ of $f(t)$ and have established the generalised Fourier representation. It should be noted that this concept just follows from the assumption of the existence of the *auto-correlation function* of $f(t)$ defined by

$$\lim_{T\to\infty}\frac{1}{2T}\int_{-T}^{T}f(t+\tau)f^*(t)\,\mathrm{d}t,$$

which has been denoted by $\varphi(\tau)$ (cf. (7.1)). This auto-correlation function $\varphi(\tau)$ is found by a summation of contributions of the different frequencies of the spectrum. This leads to consider the function $f(t)$ as a realisation of a stochastic process, consisting of a superposition of different harmonic waves, the amplitudes of which are random variables of ω.

We first remark that if $f(t)$ is a stationary process of second order,[1] with the understanding that the process is a centred process, i.e., $E\{f(t)\} = 0$, there exists a covariance function $E\{f(t+\tau)f^*(t)\}$ which is independent of t for a stationary process. If the process is *ergodic*, then this covariance function is equal to the auto-correlation $\varphi(t)$, i.e.,

$$E\{f(t+\tau)f^*(t)\} = \lim_{T\to\infty} \frac{1}{2T}\int_{-T}^{T} f(t+\tau)f^*(t)\,\mathrm{d}t. \tag{7.8}$$

This means that from the relation (7.7), we can consider the process $f(t)$ as a superposition of the stochastic process $A(\omega)$, which consists of functions of the frequency ω with independent increments in such a way that $E\{\mathrm{d}A(\omega)\,\mathrm{d}A^*(\omega)\} = 0$, if $\mathrm{d}\omega \neq \mathrm{d}\omega'$. Thus, if we put $\mathrm{d}F(\omega) = 2E\{\mathrm{d}A(\omega)\,\mathrm{d}A^*(\omega)\}$, we can express $f(t)$ as

$$f(t) = \int_{-\infty}^{\infty} \mathrm{e}^{\mathrm{i}\omega t}\,\mathrm{d}A(\omega). \tag{7.9}$$

In the case of surface waves, we have the vector $\boldsymbol{x} = (x, y)$ and the representation takes the form of a superposition of plane waves as a stochastic integral,

$$\eta(x, y, t) = \iint_{-\infty}^{\infty} \mathrm{e}^{\mathrm{i}(\boldsymbol{k}\cdot\boldsymbol{x}-\mathrm{i}\omega t)}\,\mathrm{d}A(\boldsymbol{k}), \tag{7.10}$$

where ω and $\boldsymbol{k} = (k_1, k_2)$ are related by the dispersion relation $\omega = H(\boldsymbol{k})$. The energy spectrum is given by $\mathrm{d}F = E\{\mathrm{d}A(\boldsymbol{k})\,\mathrm{d}A^*(\boldsymbol{k})\}$. The energy spectrum depends on both the wave number $k = \sqrt{k_1^2 + k_2^2}$ and the direction ϑ such that $k_1 = k\cos\vartheta$ and $k_1 = k\sin\vartheta$. We denote it by $\mathrm{d}F(k, \vartheta)$.

In Sect. 3.2.4 we considered the response of a ship to this incoming wave (7.10). We will see that for a purely sinusoidal incoming wave, the ship motion is also purely sinusoidal in its six degrees of freedom and there exists a transfer matrix $\mu = (\mu_k \mu_j^*)$ with entries $\mu_k \mu_j^*$ depending on ω (or k) such that $\alpha_k = \mu_k A$, where α_k is defined as the amplitude of the ship's motion in the corresponding degree of freedom. If $\mathrm{d}A(\omega)$ is a stochastic input function, then the α_k will also be stochastic. From the energy spectrum $\mathrm{d}F = E\{\mathrm{d}A\,\mathrm{d}A^*\}$ of the incoming wave, we can form the covariance matrix of the motion of the ship, $E\{\mathrm{d}\alpha_k\,\mathrm{d}\alpha^* j\}$. The diagonal elements

[1] A stochastic process $X(t)$ is said to be of second order if $E\{X^2(t)\} < \infty$ for all t in its domain of definition; the covariance function $X(t)$ is defined by $E\{[X(\tau) - m(\tau)][X(t) - m(t)]^*\}$, where $m(t) = E\{X(t)\}$.

are called the *quadrature spectrum* of the corresponding motions; the non-diagonal elements form the *cross spectrum*. It is easy to see that

$$E\{\mathrm{d}\alpha_k\,\mathrm{d}\alpha_j^*\} = E\{\mu_k\,\mathrm{d}A\,\mu_j^*\,\mathrm{d}A^*\} = \mu_k\mu_j^* E\{\mathrm{d}A\,\mathrm{d}A^*\} = \mu_k\mu_j^*\,\mathrm{d}F. \tag{7.11}$$

The matrix $\mu_k\mu_j^*$ forms the transfer matrix of the ship and is related to the α_k's by the above expression (7.11). Most experiments determine this transfer matrix from measurements of the ship motion with given incoming waves.

7.2 Shallow Water Theory

In this section we shall discuss the shallow water theory which gives a different kind of approximation from the foregoing linear theory of small amplitude. Here it is assumed that the depth of the water is sufficiently small compared with some other characteristic length associated with the horizontal direction.

For convenience, we denote by $(\bar{x}, \bar{y}, \bar{z})$ and $\bar{t}$ the dimensional space variables and time variable, respectively. The disposition of the coordinate axes is taken in the usual manner, with the $\bar{x}, \bar{z}$-plane, the undisturbed water surface and the $\bar{y}$-axis positive upward. The free surface is given by $\bar{y} = \bar{\eta}(\bar{x}, \bar{z}, \bar{t})$ and the flat bottom is given by $\bar{y} = -h$, $h > 0$ (see Fig. 7.1). The velocity components are denoted by $\bar{u}(\bar{x}, \bar{y}, \bar{z}, \bar{t})$, $\bar{v}(\bar{x}, \bar{y}, \bar{z}, \bar{t})$, $\bar{w}(\bar{x}, \bar{y}, \bar{z}, \bar{t})$ and the pressure is denoted by $\bar{p}(\bar{x}, \bar{y}, \bar{z}, \bar{t})$. We recapitulate the basic equations and boundary conditions in terms of the Euler variables $(\bar{x}, \bar{y}, \bar{z})$ and $\bar{t}$. The equations of motion in Sect. 1.1 take the form,

$$\begin{aligned}
\bar{u}_{\bar{t}} + \bar{u}\bar{u}_{\bar{x}} + \bar{v}\bar{u}_{\bar{y}} + \bar{w}\bar{u}_{\bar{z}} &= -\frac{1}{\rho}\bar{p}_{\bar{x}}, \\
\bar{v}_{\bar{t}} + \bar{u}\bar{v}_{\bar{x}} + \bar{v}\bar{v}_{\bar{y}} + \bar{w}\bar{v}_{\bar{z}} &= -\frac{1}{\rho}\bar{p}_{\bar{y}} - g, \\
\bar{w}_{\bar{t}} + \bar{u}\bar{w}_{\bar{x}} + \bar{v}\bar{w}_{\bar{y}} + \bar{w}\bar{w}_{\bar{z}} &= -\frac{1}{\rho}\bar{p}_{\bar{z}}.
\end{aligned} \tag{7.12}$$

The equation of continuity is

$$\bar{u}_{\bar{x}} + \bar{v}_{\bar{y}} + \bar{w}_{\bar{z}} = 0 \tag{7.13}$$

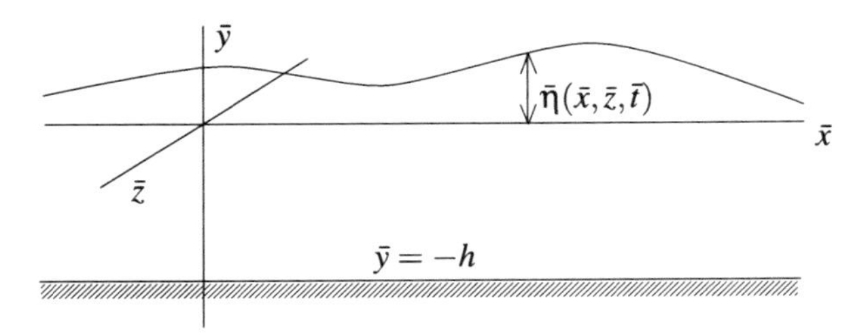

Fig. 7.1 Shallow water

and the condition of irrotational flow yields

$$\begin{aligned}\bar{u}_{\bar{y}} - \bar{v}_{\bar{x}} &= 0,\\ \bar{v}_{\bar{z}} - \bar{w}_{\bar{y}} &= 0,\\ \bar{w}_{\bar{x}} - \bar{u}_{\bar{z}} &= 0.\end{aligned} \tag{7.14}$$

At the free surface $\bar{y} = \bar{\eta}(\bar{x}, \bar{z}, \bar{t})$, we have the kinematic condition from (1.11),

$$\bar{v} = \bar{\eta}_{\bar{t}} + \bar{u}\bar{\eta}_{\bar{x}} + \bar{w}\bar{\eta}_{\bar{z}}, \tag{7.15}$$

and the condition of constant pressure from (1.13),

$$\varphi_{\bar{t}} + \frac{1}{2}(\bar{u}^2 + \bar{v}^2 + \bar{w}^2) + g\bar{\eta} = \text{constant} \tag{7.16}$$

where φ is the velocity potential which exists because of condition (7.14). By assuming the bottom flat, we have the boundary condition

$$\bar{v} = 0, \quad \text{at } \bar{y} = -h. \tag{7.17}$$

We now introduce dimensionless space variables

$$x = \frac{\bar{x}}{L}, \qquad z = \frac{\bar{z}}{L}, \qquad y = \frac{\bar{y}}{h} = \frac{\bar{y}}{\varepsilon L},$$

where $\varepsilon = h/L$ and L is a reference length in the $\bar{x}$ and $\bar{z}$ directions. Then by introducing a reference velocity U (which will be chosen later) and putting

$$u = \frac{\bar{u}}{U}, \qquad \varepsilon v = \frac{\bar{v}}{U}, \qquad w = \frac{\bar{w}}{U},$$

the equation of continuity (7.13) reads

$$u_x + v_y + w_z = 0, \tag{7.18}$$

and the vorticity free conditions (7.14) give

$$\begin{aligned}u_y - \varepsilon^2 v_x &= 0,\\ w_y - \varepsilon^2 v_z &= 0,\\ w_x - u_z &= 0.\end{aligned} \tag{7.19}$$

Taking for the time scale $t = \bar{t}U/L$, we have the equations of motion in the form

$$\begin{aligned}u_t + uu_x + vu_y + wu_z &= -\frac{1}{\rho U^2}\bar{p}_x,\\ \varepsilon^2(v_t + uv_x + vv_y + wv_z) &= -\frac{1}{\rho U^2}\bar{p}_y - \frac{\varepsilon g L}{U^2},\\ w_t + uw_x + vw_y + ww_z &= -\frac{1}{\rho U^2}\bar{p}_z.\end{aligned} \tag{7.20}$$

We now choose U such that $\varepsilon g L = U^2$. Then the above equations become, with $\Pi = \bar{p}/\rho U^2$,

$$
\begin{aligned}
u_t + u u_x + v u_y + w u_z &= -\Pi_x, \\
\varepsilon^2 (v_t + u v_x + v v_y + w v_z) &= -\Pi_y - 1, \\
w_t + u w_x + v w_y + w w_z &= -\Pi_z.
\end{aligned}
\tag{7.21}
$$

Similarly introducing the dimensionless variables $\phi = \varphi/UL$, $\eta = \bar{\eta}/h$ in (7.15) to (7.17) yields

$$
\left.
\begin{aligned}
&v = \eta_t + u\eta_x + w\eta_z \\
&\phi_t + \tfrac{1}{2}(u^2 + \varepsilon^2 v^2 + w^2) + \eta = \text{constant}
\end{aligned}
\right\} \quad \text{at } y = \eta,
\tag{7.22}
$$

$$
v = 0, \quad \text{at } y = -1.
\tag{7.23}
$$

Apparently the relevant parameter is ε^2, hence the object now is to consider solutions of (7.18)–(7.20) together with (7.22)–(7.23), depending on this small parameter and then develop in powers of ε^2.

First we assume power series developments for u, v, w and Π:

$$
\begin{aligned}
u &= u_0 + \varepsilon^2 u_1 + \varepsilon^4 u_2 + \cdots, \\
v &= v_0 + \varepsilon^2 v_1 + \varepsilon^4 v_2 + \cdots, \\
w &= w_0 + \varepsilon^2 u w_1 + \varepsilon^4 w_2 + \cdots, \\
\Pi &= u_0 + \varepsilon^2 \Pi_1 + \varepsilon^4 \Pi_2 + \cdots
\end{aligned}
\tag{7.24}
$$

and substitute them into (7.18) to (7.20). Equating to zero the coefficients of like powers of ε^2 gives the following successive system of equations:

$$
u_{0x} + v_{0y} + w_{0z} = 0,
\tag{7.25}
$$

$$
u_{0y} = 0; \qquad w_{0y} = 0; \qquad w_{0x} - u_{0z} = 0,
\tag{7.26}
$$

$$
\begin{aligned}
&u_{0t} + u_0 u_{0x} + v_0 u_{0y} + w_0 u_{0z} = -\Pi_{0x}, \\
&0 = -\Pi_{0y} - 1, \\
&w_{0t} + u_0 w_{0x} + v_0 w_{0y} + w_0 w_{0z} = -\Pi_{0z}.
\end{aligned}
\tag{7.27}
$$

Putting $\eta = \eta_0 + \varepsilon^2 \eta_1 + \varepsilon^4 \eta_2 + \cdots$, the free surface conditions are

$$
v_0 = \eta_{0t} + u_0 \eta_{0x} + w_0 \eta_{0z}
\tag{7.28}
$$

together with

$$
\Pi_0(\eta) = 0.
\tag{7.29}
$$

At the bottom we have the boundary condition

$$
v_0 = 0, \quad \text{at } y = -1,
\tag{7.30}
$$

and from the continuity equation (7.25) we then find

$$v_0(\eta) = -\int_{-1}^{\eta} (u_{0x} + w_{0z})\,\mathrm{d}y,$$

but from (7.26), since u_0 and w_0 are independent of y, it follows that

$$\begin{aligned} v_0(\eta) &= -(u_{0x} + w_{0z})(\eta + 1) \\ &= -(u_{0x} + w_{0z})(\eta_0 + 1) + \mathrm{O}(\varepsilon^2). \end{aligned} \tag{7.31}$$

Also from the second equation of (7.27), $\Pi_{0y} = -1$ and the fact that $\Pi_0(\eta) = 0$ from (7.29), we see that to the same order of approximation

$$\Pi_0 = \eta_0 - y, \tag{7.32}$$

which corresponds to the hydrostatic pressure distribution. Collecting results (7.31) and (7.32), we obtain from (7.27) the equations

$$\begin{aligned} u_{0t} + u_0 u_{0x} + w_0 u_{0z} + \eta_{0x} &= 0, \\ w_{0t} + u_0 w_{0x} + w_0 w_{0z} + \eta_{0z} &= 0, \end{aligned} \tag{7.33}$$

together with

$$\eta_{0t} + u_0 \eta_{0x} + w_0 \eta_{0z} + (\eta_0 + 1)(u_{0x} + w_{0z}) = 0,$$

from (7.28). Observe that the terms $v_0 u_{0y}$ and $v_0 w_{0y}$ in (7.27) vanish because of (7.27). Hence we arrive at, in the first approximation, the following set of quasi-linear equations for u, w and the total thickness of the water layer η:

$$\begin{aligned} u_t + u u_x + w u_z + \eta_x &= 0, \\ w_t + u w_x + w w_z + \eta_z &= 0, \\ \eta_t + u \eta_x + w \eta_z + (\eta + 1)(u_x + w_z) &= 0. \end{aligned} \tag{7.34}$$

These are the dimensionless forms of the basic equations of shallow water theory.

In the following we shall consider two special cases of (7.34) which are of importance: the steady case from (7.26) and (7.34),

$$\begin{aligned} w_x - u_z &= 0, \\ u u_x + w u_z + \eta_x &= 0, \\ u w_x + w w_z + \eta_z &= 0, \\ u \eta_x + w \eta_z + (\eta + 1) u_x + (\eta + 1) w_z &= 0, \end{aligned} \tag{7.35}$$

and the one-dimensional non-steady case from (7.34),

$$\begin{aligned} u_t + u u_x + \eta_x &= 0, \\ \eta_t + u \eta_x + (\eta + 1) u_x &= 0. \end{aligned} \tag{7.36}$$

In the first case, from the first three equations of (7.35), we find that

$$\frac{\partial}{\partial x}\left\{\frac{1}{2}(u^2+w^2)+\eta\right\}=0,$$

$$\frac{\partial}{\partial z}\left\{\frac{1}{2}(u^2+w^2)+\eta\right\}=0,$$

or

$$\eta=\frac{1}{2}\left\{(c_0-2)-(u^2+w^2)\right\},\quad c_0=\text{constant}. \tag{7.37}$$

Substitution of (7.37) into the last equation of (7.35) gives a second-order equation for the potential function ϕ:

$$\frac{\partial}{\partial x}\{[c_0-(u^2+w^2)]\phi_x\}+\frac{\partial}{\partial z}\{[c_0-(u^2+w^2)]\phi_z\}=0,$$

or

$$[c_0-(3u^2+w^2)]\phi_{xx}-4uw\phi_{xz}+[c_0-(u^2+3w^2)]\phi_{zz}=0. \tag{7.38}$$

The steady flow case is completely analogous to the steady two-dimensional gas flow.

In the second case, the unsteady one-dimensional case, (7.36) can be solved by the method of characteristics. We define here the characteristic directions α and β as those for which the linear combination of the equations yields differentiation of the functions η and u in the same direction. A linear combination of (7.36) is

$$\lambda u_t+(\lambda u+\mu(\eta+1))u_x+\mu\eta_t+(\lambda+\mu u)\eta_x=0, \tag{7.39}$$

and the differentiation of u and η are in the same direction if

$$\frac{\lambda u+\mu(\eta+1)}{\lambda}=\frac{\lambda+\mu u}{\mu},$$

which gives $\lambda=\mu\sqrt{\eta+1}$ or $\lambda=-\mu\sqrt{\eta+1}$. Hence we have finally from (7.39), putting $c=\sqrt{\eta+1}$,

$$\begin{aligned} &cu_t+c(u+c)u_x+\eta_t+(u+c)\eta_x=0,\\ &-cu_t-c(u-c)u_x+\eta_t+(u-c)\eta_x=0. \end{aligned} \tag{7.40}$$

The characteristic directions are

$$\begin{aligned} \frac{\partial}{\partial\alpha}&=\frac{\partial}{\partial t}+(u+c)\frac{\partial}{\partial x},\\ \frac{\partial}{\partial\beta}&=\frac{\partial}{\partial t}+(u-c)\frac{\partial}{\partial x}, \end{aligned} \tag{7.41}$$

and (7.40) take the form

$$c\frac{\partial u}{\partial \alpha} + \frac{\partial \eta}{\partial \alpha} = 0 \quad \text{and} \quad -c\frac{\partial u}{\partial \beta} + \frac{\partial \eta}{\partial \beta} = 0$$

or

$$\frac{\partial u}{\partial \alpha} + 2\frac{\sqrt{\eta+1}}{\partial \alpha} = 0 \quad \text{and} \quad -\frac{\partial u}{\partial \beta} + 2\frac{\sqrt{\eta+1}}{\partial \beta} = 0 \tag{7.42}$$

by making use of $c = \sqrt{\eta + 1}$. This gives the Riemann invariants

$$\begin{aligned} u + 2\sqrt{\eta+1} &= f(\beta), \\ -u + 2\sqrt{\eta+1} &= g(\alpha). \end{aligned} \tag{7.43}$$

An application of the above theory to the breaking of a dam can be found in Stoker's book [17] on water waves.

7.3 Non-linear Dispersive Waves

As we have seen in the previous section, we can solve the shallow water equations, in the first approximation, by the method of characteristics. If these characteristics intersect, a shock will develop. For this reason, it has an advantage to consider the second approximation to the equations.

We shall derive the second-order approximation for the one-dimensional flow by a method which differs slightly from the method used in the previous section, but which is somewhat shorter.

Again, we denote the dimensional coordinates system by $(\bar{x}, \bar{y})$, but for convenience, we replace the $\bar{x}$-axis on the flat bottom as shown in Fig. 7.2. The boundary value problem for φ consists of the equations

$$\begin{aligned} &\varphi_{\bar{x}\bar{x}} + \varphi_{\bar{y}\bar{y}} = 0, \quad \text{for} \begin{cases} -\infty < \bar{x} < \infty, \\ 0 < \bar{y} < \bar{\eta}(\bar{x}, \bar{t}), \end{cases} \\ &\varphi_{\bar{y}} = 0, \quad \text{at } \bar{y} = 0, \end{aligned} \tag{7.44}$$

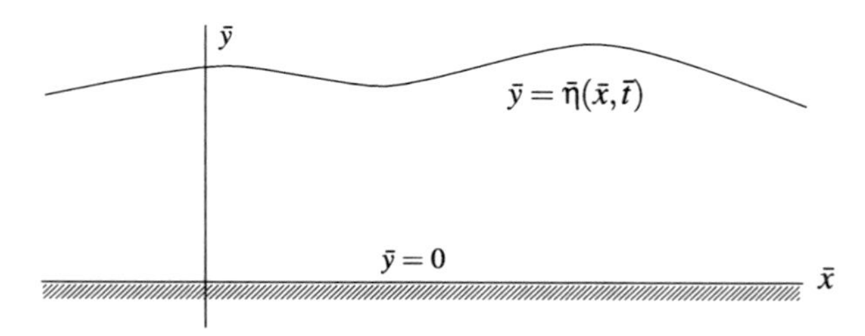

Fig. 7.2 Shallow water coordinates

and the following conditions on the free surface $\bar{y} = \bar{\eta}(\bar{x}, \bar{t})$. On the free surface the kinematic condition (7.15)

$$\bar{\eta}_{\bar{t}} + \bar{u}\bar{\eta}_{\bar{x}} = \bar{v}, \tag{7.45}$$

holds together with the dynamic condition of constant pressure $\bar{p}$, which take the form $\partial\bar{p}/\partial s = 0$. Here $\partial/\partial s$ denotes differentiation in the direction of the surface face. Since the tangent to the free surface has direction cosines $(1, \bar{\eta}_{\bar{x}})$, this dynamic condition on the free surface becomes

$$\frac{\partial\bar{p}}{\partial\bar{x}} + \frac{\partial\bar{p}}{\partial\bar{y}}\bar{\eta}_{\bar{x}} = 0, \tag{7.46}$$

where $\frac{\partial\bar{p}}{\partial\bar{x}}$ and $\frac{\partial\bar{p}}{\partial\bar{y}}$ follow from the equations of motion, that is

$$\begin{aligned} \bar{u}_{\bar{t}} + \bar{u}\bar{u}_{\bar{x}} + \bar{v}\bar{u}_{\bar{y}} &= -\frac{1}{\rho}\bar{p}_{\bar{x}}, \\ \bar{v}_{\bar{t}} + \bar{u}\bar{v}_{\bar{x}} + \bar{v}\bar{v}_{\bar{y}} &= -\frac{1}{\rho}\bar{p}_{\bar{y}} - g. \end{aligned} \tag{7.47}$$

Substitution of the above equations into (7.46) gives (cf. (1.18))

$$\bar{u}_{\bar{t}} + \bar{u}\bar{u}_{\bar{x}} + \bar{v}\bar{u}_{\bar{y}} + \bar{\eta}_{\bar{x}}(\bar{v}_{\bar{t}} + \bar{u}\bar{v}_{\bar{x}} + \bar{v}\bar{v}_{\bar{y}} + g) = 0. \tag{7.48}$$

In order to obtain the shallow water equations, we consider the complex velocity $\bar{w}$ defined by $\bar{w} = \bar{u} - \mathrm{i}\bar{v}$ which is an analytic function of the complex variable $\bar{x} + \mathrm{i}\bar{y}$ (or more precisely $\bar{x} + \mathrm{i}\bar{y}$ and $\bar{t}$) in the interior of the region occupied by the fluid. We will assume this regularity continues to remain valid up to and including the boundary of the region. Then expanding $\bar{w}$ into a Taylor expansion about $\bar{x}$ for fixed $\bar{t}$, we find

$$\bar{u} - \mathrm{i}\bar{v} = \bar{w}(\bar{x}) + \mathrm{i}\bar{y}\bar{w}'(\bar{x}) - \frac{\bar{y}^2}{2}\bar{w}''(\bar{x}) - \frac{\mathrm{i}\bar{y}^3}{6}\bar{w}'''(\bar{x}) + \cdots. \tag{7.49}$$

(Here for simplicity, we omit the argument $\bar{t}$ and henceforth in the sequel.) Since at the flat bottom $\bar{y} = 0$, we have

$$\bar{v} = \bar{v}_{\bar{x}} = \bar{v}_{\bar{x}\bar{x}} = \bar{v}_{\bar{x}\bar{x}\bar{x}} = \cdots = 0, \tag{7.50}$$

it follows that by differentiation of (7.49) with respect to $\bar{x}$, we obtain for $\bar{y} = 0$,

$$\begin{aligned} \bar{w}(\bar{x}) &= \bar{u}(\bar{x}, 0), \\ \bar{w}'(\bar{x}) &= \bar{u}_{\bar{x}}(\bar{x}, 0), \\ \bar{w}''(\bar{x}) &= \bar{u}_{\bar{x}\bar{x}}(\bar{x}, 0), \ldots. \end{aligned} \tag{7.51}$$

This gives us the expansions

$$\begin{aligned}\bar{u}(\bar{x},\bar{y}) &= \bar{u}(\bar{x},0) - \frac{\bar{y}^2}{2}\bar{u}_{\bar{x}\bar{x}}(\bar{x},0) + \frac{\bar{y}^4}{4!}\bar{u}_{\bar{x}\bar{x}\bar{x}\bar{x}}(\bar{x},0) + \cdots, \\ \bar{v}(\bar{x},\bar{y}) &= -\bar{y}\bar{u}_{\bar{x}}(\bar{x},0) + \frac{\bar{y}^3}{3!}\bar{u}_{\bar{x}\bar{x}\bar{x}}(\bar{x},0) - \frac{\bar{y}^5}{5!}\bar{u}_{\bar{x}\bar{x}\bar{x}\bar{x}\bar{x}}(\bar{x},0) + \cdots.\end{aligned} \tag{7.52}$$

We now introduce dimensionless coordinates by putting $x = \bar{x}/L$, $y = \bar{y}/h$, where h is the characteristic length of the water depth and L is the *wave length* in the $\bar{x}$-direction. Moreover, we put $\bar{u} = Uu$ with U being a reference velocity. The time $\bar{t}$ is made dimensionless by introduction of a time scale T such that $t = \bar{t}/T$. Putting $\delta = h/L$, we see that $\bar{v}$ is of order δ with respect to $\bar{u}$. We put $\bar{v} = \delta U v$. Furthermore, we refer to $u(x,0,t)$ as

$$u(x,0,t) = f(x,t). \tag{7.53}$$

Then from (7.52), we have the expansions

$$\begin{aligned}u(x,y,t) &= f - \frac{\delta^2}{2}y^2 f_{xx} + \frac{\delta^4}{4!}y^4 f_{xxxx} + \mathrm{O}(\delta^6), \\ v(x,y,t) &= -yf_x + \frac{\delta^2}{3!}y^2 f_{xxx} - \frac{\delta^4}{5!}y^5 f_{xxxxx} + \mathrm{O}(\delta^6).\end{aligned} \tag{7.54}$$

On the surface $y = \eta(x,t)$, $\eta = \bar{\eta}/h$, the kinematic condition (7.45) takes the form

$$\frac{h}{T}\eta_t + \frac{Uh}{L}u\eta_x = U\delta v, \tag{7.55}$$

or by setting $\delta^s = L/UT$ for some non-negative power s, it becomes

$$\delta^s \eta_t + u\eta_x = v. \tag{7.56}$$

The constant pressure condition (7.48) is transformed into

$$\delta^s u_t + uu_x + vu_y + \delta^2\eta_x(\delta^s v_t + uv_x + vv_y) + \eta_x\frac{gL\delta}{U^2} = 0. \tag{7.57}$$

We choose U such that $U^2 = gL\delta$ and substitute (7.54) for u and v into (7.56) and (7.57). Then we obtain a set of equations for f and η:

$$\begin{aligned}&\delta^s\eta_x + (f\eta_x + \eta f_x) - \frac{1}{2}\delta^2\eta^2\left(\eta_x f_{xx} + \frac{1}{3}\eta f_{xxx}\right) \\ &\quad + \frac{\delta^4}{4!}\eta^4\left(\eta_x f_{xxxx} + \frac{1}{5}\eta f_{xxxxx}\right) + \mathrm{O}(\delta^6) = 0, \\ &\delta^s\left[f_t - \delta^2\left(\frac{\eta^2}{2}f_{xxt} + \eta\eta_x f_{xt}\right) + \frac{\delta^4}{6}\eta^3\left(\eta_x f_{xxxt} + \frac{1}{4}\eta f_{xxxxt}\right)\right] + (ff_x + \eta_x) \\ &\quad + \delta^2\eta\left(\eta_x f_x^2 - \eta_x ff_{xx} + \frac{1}{2}\eta f_x f_{xx} - \frac{1}{2}\eta ff_{xxx}\right)\end{aligned} \tag{7.58}$$

$$+\delta^4\eta^3\left[\frac{\eta}{4}\left(\frac{1}{3}f_{xx}f_{xxx}-\frac{1}{2}f_x f_{xxxx}+\frac{1}{6}ff_{xxxxx}\right)\right.$$

$$\left.+\eta_x\left(\frac{1}{2}f_{xx}^2-\frac{2}{3}f_x f_{xxx}+\frac{1}{6}ff_{xxxx}\right)\right]+\mathrm{O}(\delta^6)=0. \qquad (7.59)$$

We construct asymptotic solutions of this set of equations by substituting for f and η the power series in δ^2:

$$\begin{aligned} f &= u_0+\delta^2 u_1+\delta^4 u_2+\cdots, \\ \eta &= \eta_0+\delta^2\eta_1+\delta^4\eta_2+\cdots. \end{aligned} \qquad (7.60)$$

There are two cases to be considered:

CASE 1: $s=0$

In this case the time scale T is adapted to the reference speed U by the relation $U=/L/T$. Substitution of the series (7.60) into (7.58) and (7.59) gives, after equating to zero the coefficients of like powers of δ^2, the first approximation:

$$\begin{aligned} \eta_{0t}+u_0\eta_{0x}+\eta_0 u_{0x} &= 0, \\ u_{0t}+u_0 u_{0x}+\eta_{0x} &= 0, \end{aligned} \qquad (7.61)$$

which are the shallow water equations for one-dimensional flow (7.33). The second approximation leads to the following:

$$\begin{aligned} &\eta_{0t}+\eta_0 u_{1x}+\eta_1 u_{0x}+\eta_{0x}u_1+\eta_{1x}u_0-\frac{\eta_0^2}{6}\left(\eta_0 u_{0xxx}+3\eta_{0x}u_{0xx}\right)=0, \\ &u_{1t}+\eta_{1x}+u_0 u_{1x}+u_{0x}u_1-\frac{1}{2}\eta_0^2 u_{0xxt}-\frac{\eta_0^2}{2}\left(u_0 u_{0xxx}-u_{0x}u_{0xx}\right) \\ &\quad -\eta_0\eta_{0x}u_{0xt}-\eta_0\eta_{0x}u_0 u_{0xx}+\eta_0\eta_{0x}u_{0x}^2=0. \end{aligned} \qquad (7.62)$$

We consider the special case of the first approximation due to a flow with constant velocity which we put equal to one at $y=-1$. In this case since (7.62) gives only the trivial solution, we must consider an approximation somewhat different than the ones used in (7.62). We reexamine (7.58) and (7.59) by considering the expansions

$$\begin{aligned} f &= 1+\delta^2 u_1+\delta^4 u_2+\cdots, \\ \eta &= 1+\delta^2\eta_1+\delta^4\eta_2+\cdots. \end{aligned} \qquad (7.63)$$

We keep all terms up to order δ^4 except those terms involving u_2 and η_2, and obtain

$$\begin{aligned} &\eta_{1t}+\eta_{1x}+\delta^2(u_1\eta_{1x}+\eta_1 u_{1x})-\frac{1}{6}\delta^2 u_{1xxx}=0, \\ &u_{1t}+u_{1x}+\eta_{1x}+\delta^2 u_1 u_{1x}-\frac{1}{2}\delta^2(u_{1xxt}+u_{1xxx})=0. \end{aligned} \qquad (7.64)$$

By setting $u = 1 + \delta^2 u_1$ and $\eta = 1 + \delta^2 \eta_1$ (or equivalently, $u_1 = \frac{1}{\delta^2}(u - 1)$ and $\eta_1 = \frac{1}{\delta^2}(\eta - 1)$), we obtain from (7.64) a variant of the equations of Boussinesq,

$$\begin{aligned} \eta_t + \eta u_x + u\eta_x - \frac{1}{6}\delta^2 u_{xxx} &= 0, \\ u_t + uu_x - \eta_x - \frac{1}{2}\delta^2(u_{xxt} + u_{xxx}) &= 0. \end{aligned} \tag{7.65}$$

The original form of the equations of Boussinesq can be obtained by the following transformation which is allowed in the order of approximation; we first replace u by

$$u = 1 + \delta^2\left(u_1 - \frac{1}{6}\delta^2 u_{1xx}\right), \tag{7.66}$$

and introducing an x'-coordinate system which moves with the main flow $U = 1$ $(x' = x - Ut)$, we then put $u'(x', t) = u(x' + Ut, t) - U$, $\eta'(x', t) = \eta(x' + Ut, t)$. This leads successively to the equations, after dropping the primes,

$$\begin{aligned} \eta_t + \eta u_x + u\eta_x &= 0, \\ u_t + uu_x + \eta_x - \frac{1}{3}\delta^2 u_{xxt} &= 0. \end{aligned} \tag{7.67}$$

Equations (7.67) are known as the *Boussinesq equations* for the one-dimensional shallow water waves.

CASE 2: $s = 2$

The first approximation now does not contain t, and we simply have

$$\begin{aligned} u\eta_{0x} + \eta_0 u_{0x} &= 0, \\ uu_{0x} + \eta_{0x} &= 0. \end{aligned} \tag{7.68}$$

This is solved by $u = 1$ and $\eta_0 = 1$, if we assume again the series expansion (7.63). Substitution in (7.58) and (7.59) gives the second approximation, the coefficients of δ^2,

$$\begin{aligned} \eta_{1x} + u_{1x} &= 0, \\ u_{1x} + \eta_{1x} &= 0, \end{aligned} \tag{7.69}$$

which shows that $u_1 = -\eta_1 + X(t)$ for some function $X(t)$. If we assume that the flow is undisturbed at infinity, then this implies

$$u_1 = -\eta_1 + \alpha \tag{7.70}$$

for some constant α.

Consider now the next approximation, the coefficients of δ^4:

$$\begin{aligned} &\eta_{1t} + u_1 \eta_{1x} + \eta_1 u_{1x} - \frac{1}{6} u_{1xxx} + u_{2x} + \eta_{2x} = 0, \\ &u_{1x} + u_1 u_{1x} - \frac{1}{2} u_{1xxx} + \eta_{2x} + u_{2x} = 0. \end{aligned} \tag{7.71}$$

Subtracting the two equations in order to eliminate u_2 and η_2 gives

$$\eta_{1t} - u_{1t} + u_1(\eta_{1x} - u_{1x}) + \eta_1 u_{1x} + \frac{1}{3} u_{1xxx} = 0. \tag{7.72}$$

By using the relation (7.70), we then reduce to an equation of η_1, the *Korteweg-De Vries* equation:

$$\eta_{1t} + \left(\alpha - \frac{3}{2}\eta_1\right)\eta_{1x} - \frac{1}{6}\eta_{1xxx} = 0. \tag{7.73}$$

In order to find a wave solution of the Korteweg-De Vries equation (7.73), we first change the coordinates in such a way so that the equation takes the form

$$\eta_t + (1 + \varepsilon\eta)\eta_x + \mu\eta_{xxx} = 0, \tag{7.74}$$

where ε and μ are small parameters depending on α. Introducing a large parameter K, we substitute

$$\eta(x,t) = U\big[KS(x,t), x, t\big] + \frac{1}{K} V\big[KS(x,t), x, t\big] + \mathrm{O}\left(\frac{1}{K^2}\right) \tag{7.75}$$

into (7.74) and with the abbreviation $p = KS(x,t)$, we obtain the approximation up to terms of order K:

$$(k - \omega)U_p + \varepsilon k U U_p + (\mu K^2) k^3 U_{ppp} + \frac{1}{K}\mathrm{O}(\mu K^2) = 0, \tag{7.76}$$

where we introduced

$$k = S_x(x,t), \qquad \omega = -S_t(x,t)$$

as the wave number and frequency, respectively.

Now identify K^2 so that $\mu K^2 = 1$ and notice that S is slowly varying with x and t. With this choice we neglect the term $\frac{1}{K}\mathrm{O}(\mu K^2)$ in (7.76), since it is of lower order. Then (7.76) is just an ordinary third-order differential equation for U as a function of p with coefficients which vary with x and t. This gives, after integration with respect to p,

$$(k - \omega)U + \frac{1}{2}\varepsilon k U^2 + k^3 U_{pp} = \frac{1}{2}\alpha. \tag{7.77}$$

Integrating once more after multiplication with U_p yields

$$k_3 U_p^2 = \alpha U + \beta + (\omega - k)U^2 - \frac{1}{3}\varepsilon k U^3. \tag{7.78}$$

Here the constants of integration α and β are functions of x and t.

Equation (7.78) is similar to the equation for the pendulum and it can be seen that U is a periodic function of p oscillating between the zeros of the right-hand side of (7.78). In general, this right-hand side has three zeros but we have to select those two zeros for which the right-hand side of (7.78) is positive between these two zeros.

We now perform a linear substitution,

$$U(p,x,t) = m(x,t) + A(x,t)\Theta(p), \tag{7.79}$$

which makes the zeros of the new function Θ between -1 and $+1$. Using the fact that $\Theta_p = 0$ for both $\Theta = -1$ and $\Theta = +1$, from (7.78) we can replace α and β by the constants (functions of x and t) m and A, and obtain the equation

$$k^3 \Theta_p^2 = (1 - \Theta^2)\left(\frac{1}{3}\varepsilon A k \Theta + \varepsilon k m + (k - \omega)\right). \tag{7.80}$$

The dependence of Θ on p can be given in implicit form as

$$p + \gamma = \sqrt{k^3} \int_{-1}^{\Theta} \frac{\mathrm{d}\Theta}{\sqrt{(1 - \Theta^2)(\frac{1}{3}\varepsilon A k \Theta + \varepsilon k m + (k - \omega))}}, \tag{7.81}$$

with $\gamma(x,t)$ as a shifting constant. The period of Θ can be normalised to unity by changing the scale p so that

$$\frac{1}{2} = \sqrt{k^3} \int_{-1}^{1} \frac{\mathrm{d}\Theta}{\sqrt{(1 - \Theta^2)(\frac{1}{3}\varepsilon A k \Theta + \varepsilon k m + (k - \omega))}}. \tag{7.82}$$

This gives a dispersion relation between k, ω, A and m. Observe that for $\varepsilon = 0$, we simply get the relation

$$\frac{1}{2} = \sqrt{\frac{k^3}{k - \omega}} \int_{-1}^{1} \frac{\mathrm{d}\Theta}{\sqrt{1 - \Theta^2}} = \pi \sqrt{\frac{k^3}{k - \omega}}, \tag{7.83}$$

which corresponds to the dispersion relation for the linear equation for cnoidal waves, i.e., (7.74) with $\varepsilon = 0$.

A second relation between k and ω is the conservation law for wave crests:

$$\frac{\partial \omega}{\partial x} + \frac{\partial k}{\partial t} = 0. \tag{7.84}$$

In order to find two more relations for k, ω, m, and A we return to the original equations (7.74) and substitute into the equation the expression (7.79) for $\eta x, t$, i.e.,

$$\eta(x,t) = m(x,t) + A(x,t)\Theta(p).$$

Since

$$\begin{aligned}
\eta_t &= m_t + A_t\Theta - KA\Theta_p\omega,\\
\eta_x &= m_x + A_x\Theta + KA\Theta_p k,\\
\eta_{xx} &= m_{xx} + A_{xx}\Theta + 2KA_x\Theta_p k + KA\Theta_p k_x + K^2 A\Theta_{pp}k^2,\\
\eta_{xxx} &= m_{xxx} + A_{xxx}\Theta + 3KA_{xx}\Theta_p k + 3KA_x\Theta_p k_x + KA\Theta_p k_{xx}\\
&\quad + 3K^2A_x\Theta_{pp}k^2 + 3K^2A\Theta_{pp}kk_x + K^3A\Theta_{ppp}k^3,
\end{aligned}$$

this substitution gives the coefficients of K (recall $\mu K^2 = 1$),

$$k^3A\Theta_{ppp} + A\Theta_p(k-\omega) + \varepsilon(m + A\Theta)A\Theta_p k \qquad (7.85)$$

which vanishes as well as we will now see. Recall that Θ is determined by (7.80):

$$k^3\Theta_p^2 = (1-\Theta^2)\left(\frac{1}{3}\varepsilon Ak\Theta + \varepsilon km + (k-\omega)\right).$$

By differentiating this equation with respect to p, we then obtain, with some reduction, the equation

$$k^3\Theta_{pp} = \frac{1}{6}\varepsilon Ak - \Theta(\varepsilon km + (k-\omega)) - \frac{1}{2}\varepsilon Ak\Theta^2. \qquad (7.86)$$

Differentiating it once more, we arrive at the equation

$$k^3\Theta_{ppp} = (\varepsilon km + (k-\omega))\Theta_p - Ak\varepsilon\Theta\Theta_p, \qquad (7.87)$$

which shows that, as we expected, the coefficient of K^1 (cf. (7.85)) vanishes.

We expect the solution to be valid for a large number of periods (order K). This requires that the coefficient of k^0 shall not grow after many periods. The coefficient of K^0,

$$m_t + A_t\Theta + m_x + A_x\Theta + (m + A\Theta)(m_x + A_x\Theta) + 3(A_xk^2 + Akk_x)\Theta_{pp}, \qquad (7.88)$$

will be integrated over one period and the integral is put to zero. This gives

$$\begin{aligned}
&m_t + A_t\int_0^1 \Theta\,\mathrm{d}p + m_x + A_x\int_0^1 \Theta\,\mathrm{d}p + \varepsilon m\left(m_x + A_x\int_0^1 \Theta\,\mathrm{d}p\right)\\
&\quad + \varepsilon Am_x\int_0^1 \Theta\,\mathrm{d}p + \varepsilon AA_x\int_0^1 \Theta^2\,\mathrm{d}p\\
&\quad + 3(A_xk^2 + AKK_x)(\Theta_p(1) - \Theta_p(0)) = 0. \qquad (7.89)
\end{aligned}$$

Note that the last term equals zero because of the periodicity of Θ_p. By setting $\gamma_n = \int_0^1 \Theta^n \,\mathrm{d}p$, we see that (7.88) can be written in the form

$$(m + A\gamma_1)_t + (m + A\gamma_1)_x + \varepsilon(m + A\gamma_1)_x(m + A\gamma_1) + \varepsilon A A_x(\gamma_2 - \gamma_1^2) = 0. \quad (7.90)$$

Apparently the term $m + A\gamma_1$ is the mean height of the wave over a period in p, since the period of p is 1. Thus, if we let $h = m + A\gamma_1$ be the mean wave height, we obtain the relation

$$\frac{\partial h}{\partial t} + \frac{\partial}{\partial x}\left(h + \frac{1}{2}\varepsilon h^2\right) + \frac{1}{2}\varepsilon(\gamma_2 - \gamma_1^2)\frac{\partial A^2}{\partial x} = 0. \quad (7.91)$$

The remaining relation can be found by a similar procedure. Multiplying (7.74) by η and substituting (7.79) for η into the equation, we find the coefficient of K,

$$A(m + A\Theta)(k - \omega)\Theta_p + \varepsilon(m + A\Theta)^2 Ak\Theta_p + A\Theta_{ppp}(m + A\Theta)k^3, \quad (7.92)$$

which is evidently zero from (7.87). The coefficient of K^0 is

$$\begin{aligned}&(m + A\Theta)(m_t + A_t\Theta + m_x + A_x\Theta) + \varepsilon(m + A\Theta)^2(m_x + A_x\Theta)\\ &\quad + 3(m + A\Theta)(A_x k^2 + Akk_x)\Theta_{pp}.\end{aligned}$$

Integrating this over one period gives the equation

$$\begin{aligned}&m(m_t + m_x) + m(A_t + A_x)\gamma_1 + (m_t + m_x)A\gamma_1 + A(A_t + A_x)\gamma_2\\ &\quad + \varepsilon m^2 m_x + \varepsilon(2mm_x A\gamma_1 + m^2 A_x\gamma_1) + \varepsilon A^2 A_x\gamma_3 + 2\varepsilon mAA_x\gamma_2\\ &\quad + \varepsilon A^2 m_x\gamma_2 + 3(mA_x k^2 + mAkk_x)(\Theta_p(1) - \Theta_p(0))\\ &\quad + 3A(A_x k^2 + Akk_x)\int_0^1 \Theta\Theta_{pp}\,\mathrm{d}p = 0.\end{aligned} \quad (7.93)$$

Note that from (7.86) we have

$$k^3\int_0^1 \Theta\Theta_{pp}\,\mathrm{d}p = \frac{1}{6}\varepsilon Ak\gamma_1 - (\varepsilon km + (k - \omega))\gamma_2 - \frac{1}{2}\varepsilon Ak\gamma_3.$$

Hence, with $h = m + A\gamma_1$, the mean wave height, we obtain from (7.93) the relation

$$\begin{aligned}&h(h_t + h_x) + A(A_t + A_x)(\gamma_2 - \gamma_1^2) + \varepsilon h^2 h_x + 2\varepsilon mAA_x(\gamma_2 - \gamma_1^2)\\ &\quad + \varepsilon A^2 m_x(\gamma_2 - \gamma_1^2) + \varepsilon A^2 A_x(\gamma_3 - \gamma_1^3)\\ &\quad + \frac{3A(A_x k^2 + Akk_x)}{k^3}\left[\frac{1}{6}\varepsilon Ak\gamma_1 - (\varepsilon km + (k - \omega))\gamma_2 - \frac{1}{2}\varepsilon Ak\gamma_3\right] = 0, \end{aligned} \quad (7.94)$$

where use has been made of the periodicity of Θ_p.

In this way we obtain (7.82) and a set of first-order differential equations (i.e. (7.84), (7.91) and (7.87)) for k, ω, m and A. In the general case these equations are rather difficult to solve. Hence we are led to consider the approximate expansion of solutions for small ε. First we make some remarks. For $\varepsilon = 0$ the original equation (7.74) is linear and we have dispersion relation (7.83), which is independent of A and h; hence the dependence of ω on A and h is a non-linear effect; moreover, since we have $\gamma_1 = \gamma_3 = 0$, we shall have trivial solutions from (7.91) and (7.94), if we put ε equal to zero. In order to take these facts into account, we introduce, instead of m, the quantity m_1 such that $h = \varepsilon m_1 + A\gamma_1$ is the mean wave height. Neglecting the terms of order ε^2, from (7.91) and (7.94) we then arrive at the following:

$$\begin{aligned} &\frac{\partial h}{\partial t} + \frac{\partial h}{\partial x} + \frac{1}{2}\varepsilon\gamma_2 \frac{\partial A^2}{\partial x} = 0, \\ &\frac{\partial A^2}{\partial t} + \frac{\partial}{\partial x}[(1 - 12\pi^2 k^2)A^2] = 0, \end{aligned} \tag{7.95}$$

together with the dispersion relation (7.83) which can be written as

$$\omega_0 = k - 4\pi^2 k^3. \tag{7.96}$$

Here we write ω_0 instead of ω in order to emphasise the relation corresponding to the linear problem. The second equation of (7.95) is obtained making use of (7.96); we observe that $\omega_0'(k) = 1 - 12\pi^2 k^2$. Finally, we introduce the second-order functions $H_0(x,t)$ and $E(x,t)$ by means of

$$H_0 = \varepsilon h, \qquad E = \varepsilon^2 A^2,$$

in terms of which (7.95) may be written in the form

$$\begin{aligned} &\frac{\partial H_0}{\partial t} + \frac{\partial}{\partial x}\left\{ H_0 + \frac{\gamma_2}{2} E \right\} = 0, \\ &\frac{\partial E}{\partial t} + \frac{\partial}{\partial x}\{\omega_0'(k) E\} = 0. \end{aligned} \tag{7.97}$$

The first one of the above equations may be considered as an average form of the equation of conservation of mass; the second one expresses the conservation of average energy of the wave train with a linear group velocity $\omega_0'(k) = 1 - 12\pi^2 k^3$. It is possible to derive from the dispersion relation (7.82) an asymptotic expansion of ω in terms of k, H_0 and E for small ε. The latter together with (7.84) and (7.97) can be utilised for determining ω as well as k, H_0 and E approximately.

Chapter 8
Shallow Water Ship Hydrodynamics

In this chapter we consider slender ships in shallow water; we discuss three related topics. We are, among others, interested in the influence of the bottom on the vertical motion of the ship. It turns out that a ship in shallow water experiences a certain sinkage and trim due to the bottom effect. This is of importance if one wishes to determine the required depth of harbours in such a way that a ship may enter safely. It is also of importance to know the wall effects for ships travelling in a shallow channel. Here we wish to know sinkage and trim but also the force and moment due to the interaction of the walls of the channel. The third topic we shall consider is the interaction among ships.

The topics have been investigated by E.O. Tuck [18], R.F. Beck [1] and R.W. Yeung [22, 23]. The first two are steady flow problems, while the third one is unsteady due to the difference in forward speed of the two ships. The method we use to solve these problems is related to the well-known method of matched asymptotic expansions. To obtain insight into this method, we apply it to the unsteady flow around a two-dimensional airfoil, where the introduction of the inner expansion is not commonly done. The incoming flow is in principle nearly parallel to the airfoil while the fluid domain is extended toward infinity. Thereafter, we treat the case of a cross-current flow around a slender body wedged between two parallel plates. The notion *blockage* will be introduced in this way. After the preliminary investigations, we shall pay attention to the three problems mentioned above.

8.1 Thin Airfoil Theory

We consider the two-dimensional thin airfoil moving with a constant velocity $U\mathbf{i}$ in an incompressible medium, see Fig. 8.1. With regard to a system of reference, fixed to the airfoil, the x-axis extending along the surface(approximately), the resulting motion of the fluid can be separated into the unperturbed flow with velocity U in the direction of the positive x-axis and a perturbation flow $\mathbf{v} = (u, v)$ which originates from the airfoil. The governing equations (1.2) and (1.2) are

$$u_x + v_y + w_z = 0, \tag{8.1}$$

A.J. Hermans, *Water Waves and Ship Hydrodynamics*,
DOI 10.1007/978-94-007-0096-3_8,

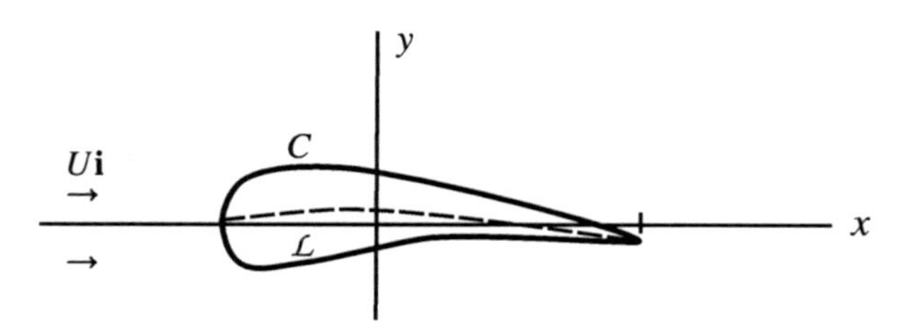

Fig. 8.1 Two-dimensional profile

$$\begin{aligned} u_t + (U+u)u_x + v u_y &= -\frac{1}{\rho} p_x, \\ v_t + (U+u)v_x + v v_y &= -\frac{1}{\rho} p_y, \end{aligned} \tag{8.2}$$

where p and ρ are pressure and density of the fluid. Because $|\mathbf{v}| = \sqrt{u^2+v^2} \ll U$, we may linearise the equation of motion (8.2), so that

$$\begin{aligned} u_t + U u_x &= -\frac{1}{\rho} p_x, \\ v_t + U v_x &= -\frac{1}{\rho} p_y. \end{aligned} \tag{8.3}$$

We now eliminate p from these two equations and introduce the vorticity vector

$$\boldsymbol{\gamma} := -\left(\frac{\partial v}{\partial x} - \frac{\partial u}{\partial y}\right)\mathbf{k} = \gamma \mathbf{k},$$

where $\mathbf{k}$ is the unit vector in the direction normal to the x, y-plane. This leads to the equation for the scalar function $\gamma = \gamma(x, y, t)$:

$$\gamma_t + U\gamma_x = 0, \tag{8.4}$$

from which we see that γ must have the form

$$\gamma = \gamma(x - Ut, y) \tag{8.5}$$

i.e., the vorticity is in linearised approximation carried along the undisturbed flow, in general along the stream lines. In the region where $\gamma = 0$, the velocity field has a potential $\tilde{\Phi}$, such that

$$\tilde{\mathbf{v}} = \operatorname{grad} \tilde{\Phi} = \nabla \tilde{\Phi},$$

and if the total field is given by

$$\tilde{\Phi} = Ux + \varphi,$$

clearly then the velocity potential φ satisfies the two-dimensional Laplace equation

$$\varphi_{xx} + \varphi_{yy} = 0. \tag{8.6}$$

Moreover, the pressure p and the potential φ are related by the linear Bernoulli equation

$$\varphi_t + U\varphi_x = -\frac{p}{\rho} + m(t), \tag{8.7}$$

where $m(t)$ is an arbitrary function of t.

To formulate the boundary value problem, we now consider a thin airfoil with the upper and lower surface given by

$$y = f(x,t) \pm g(x,t),$$

where the camber function $f(x,t) = \varepsilon F(x,t)$ and the thickness function $g(x,t) = \varepsilon G(x,t)$. Here ε is the small slenderness parameter. We denote the boundary of the airfoil by S. Then, the boundary value problem is defined by (8.6) in the exterior region to the airfoil together with the boundary condition (cf. (1.9))

$$\frac{\partial\varphi}{\partial n} = -\frac{\partial}{\partial n}(Ux),$$

or

$$\frac{\partial\varphi}{\partial y} = \varepsilon\left\{(F \pm G)_t + \left(\frac{\partial\varphi}{\partial x} + U\right)(F \pm G)_x\right\}, \quad \text{on } S, \tag{8.8}$$

and the condition at infinity is

$$\nabla\varphi \to \mathbf{0}, \quad \text{as } |\mathbf{x}| \to \infty, \tag{8.9}$$

First we consider the field in the vicinity of the airfoil where $y = O(\varepsilon)$. This can be done by a proper coordinate stretching of the coordinate

$$y = \varepsilon Y.$$

In this inner region, we denote the potential by φ^i, the inner potential while outside the region by φ^o, the outer potential. We assume that φ^i has a regular expansion in ε

$$\varphi^i \sim \sum_{k=1}^{\infty} \varepsilon^k \varphi_k^i(x, Y, t). \tag{8.10}$$

In terms of the stretched variable Y, the Laplace equation (8.6) in the near field becomes

$$\varphi^i_{YY} + \varepsilon^2 \varphi^i_{xx} = 0$$

into which by substituting (8.8), and equating like powers of ε, we obtain

$$\varphi^i_{kYY} = \begin{cases} 0, & k = 1, 2, \\ -(\varphi^i_{k-2})_{xx}, & k = 3, 4, \ldots. \end{cases} \tag{8.11}$$

The boundary condition (8.8) on the surface $Y = F \pm G$ becomes

$$\varepsilon\varphi^i_{1Y} + \varepsilon^2\varphi^i_{2Y} + \cdots = \varepsilon^2\big\{(F_t + UF_x) \pm (G_t + UG_x)\big\} + \cdots. \tag{8.12}$$

Hence from (8.11) and (8.12) it is easy to see that, for $y \lessgtr 0$,

$$\begin{aligned} \varphi^i_1(x, Y, t) &= A^{\pm}_1(x, t), \\ \varphi^i_2(x, Y, t) &= A^{\pm}_2(x, t) + \big\{(F_t + UF_x) \pm (G_t + UG_x)\big\}Y, \end{aligned} \tag{8.13}$$

where $A^{\pm}_k$ are unknowns. To determine $A^{\pm}_k$ we now use the matching principle in singular perturbation theory. We first write $y = \varepsilon Y$. This leads to

$$\varphi^i(x, y, t, \varepsilon) = \varepsilon\big[A^{\pm}_1(x, t) + \big\{(F_t + UF_x) \pm (G_t + UG_X)\big\}y\big] + \varepsilon^2 A^{\pm}_2(x, t), \tag{8.14}$$

which is supposed to be matched with the outer solution φ^o in the common region where both inner and outer solution are valid.

The outer potential φ^o has a regular perturbation in ε of the form

$$\varphi^o \sim \sum_{k=1}^{\infty} \varepsilon^k \varphi^o_k(x, y, t). \tag{8.15}$$

One term of the outer potential in terms of inner variables then reads

$$\varphi^o(x, y, t, \varepsilon) = \varepsilon\varphi^o_1(x, \varepsilon Y, t) = \varepsilon\varphi^o_1(x, 0, t) + \varepsilon^2 Y\varphi^o_{1y}(x, 0, t) + \cdots. \tag{8.16}$$

The matching condition now requires that one term of (8.14) equals two terms of (8.16) which yields

$$\begin{aligned} \varphi^o_1(x, \pm 0, t) &= A^{\pm}_1(x, t), \\ \varphi^o_{y1}(x, \pm 0, t) &= (F_t + UF_x) \pm (G_t + UG_x), \quad \text{on } \mathscr{L}, \end{aligned} \tag{8.17}$$

where $\mathscr{L}$ is the projection of the airfoil on the x-axis. This last relation is well known in linearised airfoil theory. Hence $A^{\pm}_1$ is completely determined if one has φ^o_1.

We notice that the pressure is continuous on the x-axis outside $\mathscr{L}$ and has jump discontinuities along $\mathscr{L}$ because of the potential jump $[\varphi](x, t) := \varphi(x, +0, t) - \varphi(x, -0, t)$. The fact that the pressure jump $[p] = 0$ outside $\mathscr{L}$ leads to

$$[\varphi]_t + U[\varphi]_x = 0,$$

for the potential jump $[\varphi]$ which implies that $[\varphi]$ must be a function of the variable $x - Ut$, say

$$[\varphi](x, t) = f(x - Ut). \tag{8.18}$$

This means that $[\varphi]$ propagates along lines $y=0$. In front of the airfoil, for $x=-\infty$ we have $[\varphi]=0$. In the wake behind the foil the value of $[\varphi]$ differs from zero. If at the trailing edge x_T, $[\varphi]$ is known to be

$$[\varphi]=h(x_T,t),$$

then we have from (8.18):

$$[\varphi](x,t)=h\left(x_T,t-\frac{x-x_T}{U}\right). \tag{8.19}$$

Only if $[\varphi]$ is specified in the wake, then the boundary value problem for the potential φ^o may be stated properly as we will see.

The formulation of the boundary condition (8.15) suggests a splitting of the outer potential in an even and an odd part,

$$\varphi^o(x,y,t)=\phi_e+\phi_o, \tag{8.20}$$

where

$$\begin{aligned}\phi_e(x,y,t)&=\phi_e(x,-y,t), \quad \text{and}\\ \phi_o(x,y,t)&=-\phi_o(x,-y,t),\end{aligned}$$

from which we obtain odd and even derivatives, respectively:

$$\begin{aligned}\frac{\partial\phi_e}{\partial y}(x,y,t)&=-\frac{\partial\phi_e}{\partial y}(x,-y,t),\\ \frac{\partial\phi_o}{\partial y}(x,y,t)&=\frac{\partial\phi_o}{\partial y}(x,-y,t).\end{aligned} \tag{8.21}$$

We now consider, for $y\geq 0$,

$$\begin{aligned}\varphi_o(x,y,t)&=\phi_e(x,y,t)+\phi_o(x,y,t),\\ \varphi_o(x,-y,t)&=\phi_e(x,y,t)-\phi_o(x,y,t).\end{aligned} \tag{8.22}$$

The matching condition (8.17) implies

$$\left.\begin{aligned}\frac{\partial\phi_e}{\partial y}&=\left(\frac{\partial}{\partial t}+U\frac{\partial}{\partial x}\right)G\\ \frac{\partial\phi_o}{\partial y}&=\left(\frac{\partial}{\partial t}+U\frac{\partial}{\partial x}\right)F\end{aligned}\right\} \quad \text{at } y=0. \tag{8.23}$$

Across the x-axis outside $\mathscr{F}$, ϕ_e is continuous but ϕ_o can be discontinuous with

$$\phi_o(x,+0,t)=-\phi_o(x,-0,t). \tag{8.24}$$

The discontinuity in the wake for the velocity potential φ^o gives a condition for the odd part ϕ_o. Hence the boundary value problem for φ^o splits up in two problems, one pertaining to the thickness distribution for the function ϕ_e, the other one pertaining to the camber distribution for the function ϕ_o.

To solve the boundary value problems for ϕ_e and ϕ_o it is convenient to use the fact that both functions are solutions of the two-dimensional Laplace equation. Hence the use of the analysis of complex functions in the $z = x + \mathrm{i}y$ plane is desirable. In particular, we introduce

$$\chi(z,t) = \phi + \mathrm{i}\psi,$$

with ψ being the stream function and ϕ the potential function. Here χ is an analytic function in the z-plane except on a part of the x-axis. Because of the structure of the boundary conditions on the x-axis, it is convenient to consider the complex velocity

$$W(z,t) := \frac{\mathrm{d}\chi}{\mathrm{d}z} = u - \mathrm{i}v. \tag{8.25}$$

The even part We determine a function W_e which is analytic in the complex z-plane except on the part $\mathscr{L}$ of the x-axis. We take the leading edge at $x = -1$ and the trailing edge at $x = -1$. On the segment $-1 \le x \le 1$, $y = 0$, we have the relation

$$W_e^+ - W_e^- = -2\mathrm{i}\left(\frac{\partial}{\partial t} + U\frac{\partial}{\partial x}\right)G,$$

where $W^\pm$ denotes $W(x \pm \mathrm{i}0, t)$, while for $|x| > 1$, $y = 0$,

$$W_e^+ - W_e^- = 0.$$

The solution $W_e(z,t)$ can be represented in the form

$$W_e(z,t) = -\frac{1}{\pi}\int_{-1}^{1} \frac{(\frac{\partial}{\partial t} + U\frac{\partial}{\partial \xi})G}{\xi - z}\,\mathrm{d}\xi, \tag{8.26}$$

because the function $W_e(z,t)$ has to go to zero at infinity.

The velocity potential can be derived from (8.26). We notice that

$$\chi_e(z,t) = \frac{1}{\pi}\int_{-1}^{1}\left[\left(\frac{\partial}{\partial t} + U\frac{\partial}{\partial \xi}\right)G\right]\log(\xi - z)\,\mathrm{d}\xi, \tag{8.27}$$

from which follows that

$$\phi_e(x,y,z) = \Re\chi_e(z,t) = \frac{1}{2\pi}\int_{-1}^{1}\sigma(\xi,t)\log\left(\sqrt{(x-\xi)^2 + y^2}\right)\mathrm{d}\xi. \tag{8.28}$$

Here

$$\sigma(\xi,t) = 2\left(\frac{\partial}{\partial t} + U\frac{\partial}{\partial \xi}\right)G(\xi,t),$$

is the source distribution.

The odd part Next we determine a function $W_o(z,t)$ which is analytic in the complex z-plane except on $\mathscr{L}$ and $\mathscr{W}$ of the x-axis, where $\mathscr{W}$ is the downstream part of the axis (the wake). For $x < -1$, $y = 0$ we have $W_o^+(z,t) - W_o^-(z,t) = 0$ because the velocity is continuous. For $-1 \le x \le 1$, $y = 0$ we have

$$W_o^+ + W_o^- = -2\mathrm{i}\left(\frac{\partial}{\partial t} + U\frac{\partial}{\partial x}\right)F,$$

while for $x > 1$,

$$W_o^+ - W_o^- = [\phi_{ox}] = u^+ - u^-,$$

which follows from (8.19).

To find $W_o(z,t)$, we use the theory of singular integral equations (see Sect. 9.3), and represent $W_o(z,t)$ by the Cauchy integral

$$W_o(z,t) = -\frac{1}{2\pi\mathrm{i}}\int_{\mathscr{L}+\mathscr{W}} \frac{\gamma(\xi,t)}{\xi - z}\,\mathrm{d}\xi, \tag{8.29}$$

where $\gamma(\xi,t)$ is a real valued function which is equivalent to a vortex distribution to be determined. By applying the Plemelj formulae to (8.29) we obtain:

$$\begin{aligned} W_o^+(\xi_0,t) - W_o^-(\xi_0,t) &= -\gamma(\xi_0,t), \\ W_o^+(\xi_0,t) - W_o^+(\xi_0,t) &= -\frac{1}{\pi\mathrm{i}}⨍_{\mathscr{L}+\mathscr{W}} \frac{\gamma(\xi,t)}{\xi - \xi_0}\,\mathrm{d}\xi, \end{aligned} \tag{8.30}$$

where we use $⨍$ to denote the principal value of the integral. From this it follows that

$$\gamma(\xi_0,t) = -[u](\xi_0,t), \quad \text{for } \xi_0 > -1,$$

and for $\xi_0 > 1$, we know that (8.5) holds,

$$\gamma(\xi_0,t) = \gamma(\xi_0 - Ut). \tag{8.31}$$

If we consider the situation at $t = 0$ when the total circulation equals zero, then we know that the total circulation remains zero. Hence we have

$$\int_{-1}^{\infty} \gamma(\xi,t)\,\mathrm{d}\xi = \int_{\mathscr{L}} \gamma(\xi,t)\,\mathrm{d}\xi + \int_{\mathscr{W}} \gamma(\xi,t)\,\mathrm{d}\xi = \Gamma(t) + \int_{1}^{\infty} \gamma(\xi,t)\,\mathrm{d}\xi = 0, \tag{8.32}$$

where $\Gamma(t)$ is the circulation around the foil.

If we differentiate (8.32) with respect to t and use (8.31), we obtain

$$\frac{\mathrm{d}\Gamma}{\mathrm{d}t} = -U\gamma(1,t). \tag{8.33}$$

If we start with a steady motion at $t = 0$, the vortex strength in the wake can be calculated easily by a step by step integration because of the relations (8.33) and (8.31).

To assure that (8.30) has a unique solution for the unknown $\gamma(\xi, t)$, we furthermore require the Kutta condition which requires finite velocity at the trailing edge, which means that the pressure jump along the x-axis is a continuous function. Therefore we require

$$\gamma(1+0, t) = \gamma(1-0, t). \tag{8.34}$$

From (8.30) it follows for $|x| < 1$ that we have to solve a singular integral equation

$$W_o^+ + W_o^- = -2\mathrm{i}v(x, t) = -\frac{1}{\pi \mathrm{i}} \fint_{-1}^{+1} \frac{\gamma(\xi, t)}{\xi - x} \,\mathrm{d}\xi - \frac{1}{\pi \mathrm{i}} \int_1^\infty \frac{\gamma(\xi, t)}{\xi - x} \,\mathrm{d}\xi, \tag{8.35}$$

where we consider $\gamma(\xi, t)$ to be known for $\xi > 1$. Hence the integral equation becomes

$$\frac{1}{\pi \mathrm{i}} \fint_{-1}^{+1} \frac{\gamma(\xi, t)}{\xi - x} \,\mathrm{d}\xi = \lambda(x, t). \tag{8.36}$$

Here $\lambda(x, t)$ is defined by

$$-\mathrm{i}\lambda(x, t) := 2v(x, t) + \frac{1}{\pi} \int_1^\infty \frac{\gamma(\xi - Ut)}{\xi - x} \,\mathrm{d}\xi$$

which can be computed stepwise.

By introducing

$$\Lambda(z, t) = \frac{1}{2\pi \mathrm{i}} \int_{-1}^{1} \frac{\gamma(\xi, t)}{\xi - z} \,\mathrm{d}\xi, \tag{8.37}$$

the integral equation (8.36) can be solved by using the Plemelj formulas. We determine $\Gamma(z, t)$ such that

$$\Lambda^+(x, t) + \Lambda^-(x, t) = \lambda(x, t), \quad |x| < 1, \tag{8.38}$$

and Λ is analytic elsewhere. Clearly then

$$\gamma(x, t) = \Lambda^+(x, t) - \Lambda^-(x, t), \quad |x| < 1. \tag{8.39}$$

Observe that (8.38) is of the form (9.27) in the Sect. 9.3 with $G(s) = 1$.

For the special case when $\lambda = 0$ in (8.38), it can easily be found that

$$\Lambda_0(z, t) = \frac{1}{\sqrt{1 - z^2}},$$

is a solution of (8.38), where the branch cut is chosen between $-1 < x < 1$. From this particular solution Λ_0, it follows that the solution of (8.38) admits the representation (see (9.28))

$$\Lambda(z, t) = \frac{\Lambda_0(z, t)}{2\pi \mathrm{i}} \left\{ \int_{-1}^{+1} \frac{\lambda(\xi, t)}{\Lambda_0^+(\xi, t)} \frac{1}{\xi - z} \,\mathrm{d}\xi + P(z, t) \right\}, \tag{8.40}$$

where $P(z,t)$ is an entire function of z. Consequently, by using the definition of Λ_0, the vortex strength becomes

$$\gamma(x,t) = \frac{1}{\pi \mathrm{i}\sqrt{1-x^2}}\left\{ \fint_{-1}^{+1} \frac{\lambda(\xi,t)}{\xi - x}\sqrt{1-\xi^2}\,\mathrm{d}\xi + P(x,t)\right\} \tag{8.41}$$

where

$$\lambda(\xi,t) = \mathrm{i}\left\{ 2v(\xi,t) + \frac{1}{\pi}\int_1^{\infty} \frac{\gamma(s-Ut)}{s-\xi}\,\mathrm{d}s, \right\}$$

and

$$v(\xi,t) = \left(\frac{\partial}{\partial t} + U\frac{\partial}{\partial x}\right) F(\xi,t).$$

Because of the behaviour of the field at infinity, $P(x,t)$ has to be a constant with respect to x, $P(t)$. If we integrate (8.41) with respect to x along $\mathscr{L}$ we find

$$P(t) = \mathrm{i}\Gamma(t) = \mathrm{i}\int_{-1}^{+1} \gamma(x,t)\,\mathrm{d}x. \tag{8.42}$$

If we start from the situation where the flow is symmetric as in the case of passing ships which we shall treat later on, the vortex distribution can be calculated. In that case $v_o(x,t) = 0$ in (8.35) for $t < 0$ and $v_o(x,t)$ is given for $t \geq 0$ and by means of the time step procedure $\gamma(x,t)$ can be determined uniquely from (8.5), (8.33), (8.34), (8.41)and (8.42). If the problem is not given as an initial value problem, then a more general procedure has to be followed.

We may now compute the lift force and moment acting on the airfoil. From Bernoulli's equation (8.7), we obtain the lift force as

$$\mathscr{L} = \rho\int_{-1}^{+1}\left\{\frac{\partial}{\partial t}[\phi_o] + U\frac{\partial}{\partial x}[\phi_o]\right\}\mathrm{d}x = -\rho\int_{-1}^{+1}\left\{\frac{\partial}{\partial t}\int^{x}\gamma\,\mathrm{d}s + U\gamma\right\}\mathrm{d}x, \tag{8.43}$$

and the moment about $x = 0$ becomes

$$\mathscr{M} = \rho\int_{-1}^{+1}\left\{\frac{\partial}{\partial t}[\phi_o] + U\frac{\partial}{\partial x}[\phi_o]\right\}x\,\mathrm{d}x = -\rho\int_{-1}^{+1}\left\{\frac{\partial}{\partial t}\int^{x}\gamma\,\mathrm{d}s + U\gamma\right\}x\,\mathrm{d}x. \tag{8.44}$$

Once $\gamma(x,t)$ is known along the foil, $\mathscr{L}$ and $\mathscr{M}$ may be computed easily.

8.2 Slender Body Theory

We now consider the situation where a slender body is placed between two parallel walls with a distance a (see Fig. 8.2). The uniform flow is given by $\mathbf{U} = (0, V, 0)$ as $|x| \to \infty$ with constant V. In particular, we assume that $a = \mathrm{O}(\varepsilon)$, where ε is the slenderness parameter $\varepsilon = B/L$, L the length and B the width of the slender

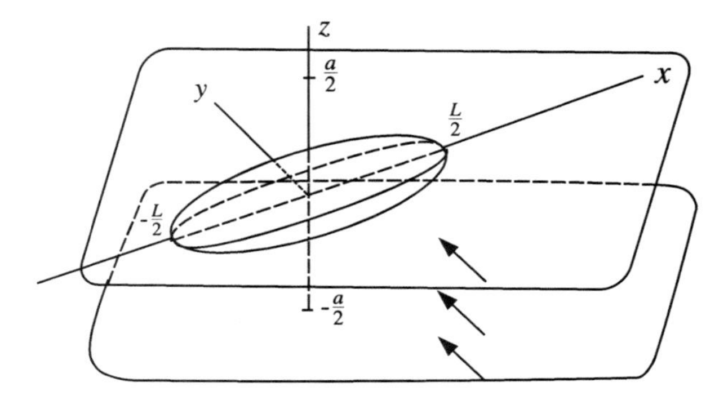

Fig. 8.2 Body between parallel walls

Fig. 8.3 The inner field

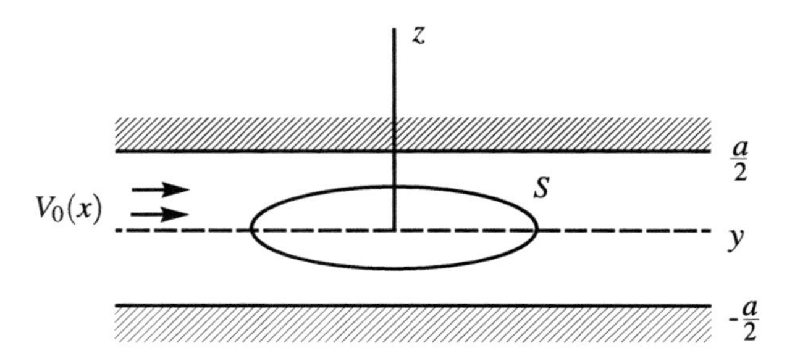

body. The region where both y and z are small will be called the inner region, see Fig. 8.3, while the region where $y = \mathrm{O}(1)$ is called the outer region. As in Sect. 8.1, we may stretch the coordinates to formulate the inner problem, which becomes a two-dimensional problem. This leads to the following boundary value problem for the first order approximation of the potential φ^i in the inner region (written in the original coordinates y, z):

$$\begin{aligned} &\varphi^i_{yy} + \varphi^i_{zz} = 0, \quad -\frac{a}{2} \le z \le \frac{a}{2}, -\infty < y < \infty, \\ &\frac{\partial \varphi^i}{\partial n} = 0, \quad \text{on the boundary of the slender body } S, \\ &\varphi^i \to V_0(x)(y \mp C(x)), \quad \text{as } y \to \pm\infty, \end{aligned} \tag{8.45}$$

where $V_0(x)$ will be determined from the matching condition later. Here $C(x)$ is referred to as the *blockage* constant. To illustrate the significance of $C(x)$, we consider the special case when a cascade of dipoles with orientation in the y-direction is superimposed on the current V_0, and these dipoles are situated at $y = 0, z = na$ with integer n. The strength of the dipoles is chosen in such a way that the flow due to one dipole in a flow V_0 describes the flow around a circular cylinder with radius ρ_1. Hence the cascade flow is the flow between two parallel walls at $z = a/2$ around a cylinder whose shape differs slightly from a circle.

The potential φ^i becomes

$$\varphi^i(y,z) = V_0(x)\left\{ y + \sum_{n=-\infty}^{\infty} \frac{\rho_1^2 y}{y^2 + (z-na)^2} \right\}.$$

For large values of $|y|$ and $|z| < a/2$, we see that

$$\varphi^i(y,z) = V_0(x)\left(y + \frac{S}{a} \right), \tag{8.46}$$

where $S = \pi\rho_1^2$ the surface area. and the blockage constant $C = S/a$. In general, a relation between the surface area S, the virtual mass m_{yy} and the blockage coefficient C can be derived such that

$$m_{yy} = -\rho S + 2\rho a C. \tag{8.47}$$

In particular, for a flat plate of length l, perpendicular to the flow between two parallel walls, m_{yy} becomes

$$m_{yy} = -\frac{2}{\pi}\rho a^2 \log\left(\cos\frac{\pi l}{2a} \right) = 2\rho a C,$$

since $S = 0$ in (8.47). Hence for most shapes the blockage coefficient may be calculated and we consider it known.

The potential in the outer region is obtained by a formal stretching of the z-coordinate. It turns out that to first and second approximation φ^o has to be a function of x and y only, and is a solution of the two-dimensional Laplace equation

$$\varphi_{xx} + \varphi_{yy} = 0,$$

the boundary conditions are

$$\varphi \sim Vy, \quad \text{as } y \to \infty,$$

and

$$\frac{\partial \varphi^o}{\partial y} = V_0(x), \quad \text{for } -1 < x < 1 \text{ and } y = 0.$$

By introducing the complex velocity function $f(\zeta) = u - \mathrm{i}v$ where $\zeta = x + \mathrm{i}y$, these conditions simplify as

$$f(x \pm \mathrm{i}0) = -\mathrm{i}V_0(x), \quad \text{for} |x| < 1,$$
$$f(\zeta) = -\mathrm{i}V \qquad |\zeta| \to \infty,$$

and $f(\zeta)$ is analytic in the complex ζ-plane such that

$$f^+(x) + f^-(x) = -2\mathrm{i}V_0(x), \quad \text{for } |x| < 1. \tag{8.48}$$

The flow around the endpoints may be singular, which leads to the choice of the solution of (8.48) for $V_0 = 0$,

$$f_0(\zeta) = (1 - \zeta^2)^{-1/2}, \tag{8.49}$$

and the solution of (8.48)

$$f(\zeta) = -\frac{f_0(\zeta)}{\pi} \int_{-1}^{1} \frac{V_0(\xi)}{f_0^+(\xi)} \frac{1}{\xi - \zeta} \, \mathrm{d}\xi + f_0(\zeta) P(\zeta), \tag{8.50}$$

where, to fulfil the condition at infinity, $P(\zeta)$ has to be of the form

$$P(\zeta) = V\zeta + P(0).$$

We now match the velocity in the x-direction of the outer and inner potentials and obtain from (8.49)

$$u(x, \pm 0) = \mp \frac{1}{\pi\sqrt{1 - x^2}} \fint_{-1}^{+1} \frac{V_0(\xi)\sqrt{1 - \xi^2}}{\xi - x} \, \mathrm{d}\xi \mp \frac{Vx + P(0)}{\sqrt{1 - x^2}} = \pm \frac{\mathrm{d}}{\mathrm{d}x}(V_0 C), \tag{8.51}$$

where $P(0)$ and $V_0(x)$ remain to be determined. To do so we integrate (8.51) with respect to x from -1 to x:

$$\begin{aligned} &V_0(x)C(x) - V_0(-1)C(-1) \\ &= -\frac{1}{\pi} \int_{-1}^{+1} V_0(\xi) K(x, \xi) \, \mathrm{d}\xi + V\sqrt{1 - x^2} - P(0)\left(\frac{\pi}{2} + \arcsin x\right), \end{aligned} \tag{8.52}$$

where $K(x, \xi)$ is a symmetric kernel of the form

$$K(x, \xi) = \fint_{-1}^{x} \frac{\sqrt{1 - \xi^2}}{\sqrt{1 - s^2}} \frac{\mathrm{d}s}{\xi - s} = \frac{1}{2} \log\left[\frac{1 - \xi x + \sqrt{1 - \xi^2}\sqrt{1 - x^2}}{1 - \xi x - \sqrt{1 - \xi^2}\sqrt{1 - x^2}}\right]. \tag{8.53}$$

For a slender body with rounded endpoints where $C(\pm 1) = 0$ it follows that $P(0) = 0$. If the cross flow problem is combined with a lateral flow, a Kutta condition may be imposed at $x = 1$. In this case it can be shown that

$$P(0) = -V \pm \frac{1}{\pi} \int_{-1}^{+1} V_0(\xi) \sqrt{\frac{1 + \xi}{1 - \xi}} \, \mathrm{d}\xi. \tag{8.54}$$

With $P(0)$ determined, (8.52) is a Fredholm integral equation of the second kind for the unknown $V_0(x)$. For large values of $C(x)$, we can find an approximate solution of (8.52) in the form

$$V_0(x)C(x) = V\sqrt{1 - x^2}, \tag{8.55}$$

in the case $P(0) = 0$, while

$$V_0(x)C(x) = V\left[\sqrt{1 - x^2} + \frac{\pi}{2} + \arcsin x\right] \tag{8.56}$$

in the latter case, where the lift force and moment may be calculated. We find, as is derived in [12],

$$\mathscr{L} = 2\rho a V_0(1)C(1),$$

and

$$\mathscr{M} = 2\rho a V_0(1)C(1) - \rho V \forall - 2\rho a \int_{-1}^{+1} V_0(x)C(x)\,\mathrm{d}x,$$

where $\forall = \int_{-1}^{+1} S\,\mathrm{d}x$ is the volume of the body; for more details see [12].

8.3 Free Surface Effects

In the preceding sections, we dealt with flow problems in an infinite fluid medium. In this section, we derive equations for shallow water ship problems where a free surface plays an important role. We consider the steady flow around a slender ship and assume that the angle of attack of the undisturbed flow is small. This situation occurs quite often near shallow water harbour entrances or a ship manoeuvring in shallow water where the variations in the course of the ship are with such a time scale that the flow may be considered to be stationary for each time interval. To be more specific, we choose the coordinate system fixed to the ship, the positive x-axis is directed in the direction of the projection of the incoming flow and the incoming flow makes an angle α with the x-axis. In contrast with the preceding chapters it is convenient to choose the positive z-axis upwards. In Fig. 8.4, B^* denotes the width of the ship, T^* the draft of the ship and L the length of the ship. The depth of the water is denoted by h^*. We assume that the velocity of the uniform flow is given by $\mathbf{U} = (U, V^*, 0)$ where U and V^* are constants. The equation for the ship hull is given by

$$y = \pm f(x, z) = \pm \varepsilon F(x, z)$$

where ε is defined by $\varepsilon = B^*/L$. The other dimensions are chosen such that $h^*/L, T^*/L$ and $V^*/U = \mathrm{O}(\varepsilon)$, as $\varepsilon \to 0$. Furthermore, the Froude number defined

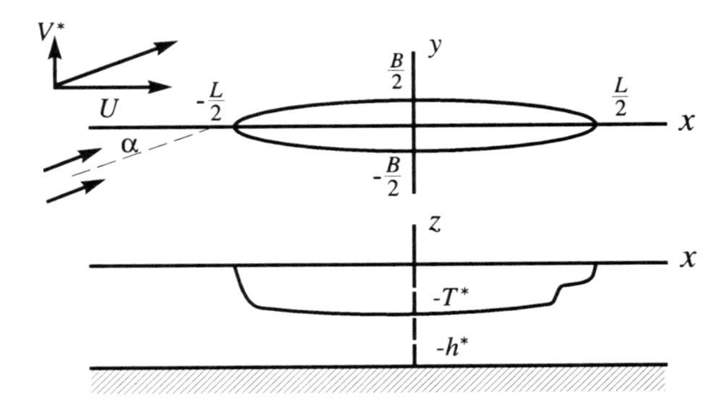

Fig. 8.4 Ship in shallow water

with the depth as characteristic length is denoted by

$$F_{h^*}^2 = \frac{|\mathbf{U}|^2}{gh^*} = \mathrm{O}(1) \quad \text{as } \varepsilon \to 0,$$

and the Froude number with respect to the length L is given by

$$F_L = \frac{|\mathbf{U}}{\sqrt{gL}} = \mathrm{O}(\varepsilon^{1/2}).$$

First we consider the outer region at a finite distance from the ship where

$$x, y = \mathrm{O}(1) \quad \text{and} \quad z = \mathrm{O}(\varepsilon).$$

Denoting the unperturbed potential by $\varphi_0^o, \varphi_0^o = Ux + V^* y$, and the perturbed potential by φ^o, we have

$$\begin{aligned} \mathbf{u} &= \nabla(\varphi_0^o + \varphi^o) =: \nabla\Phi^o, \\ \Delta\varphi^o &= 0. \end{aligned} \tag{8.57}$$

At the flat bottom $z = -h^*$, we have

$$\frac{\partial \varphi^o}{\partial z} = 0. \tag{8.58}$$

At the free surface $z = \zeta(x, y)$, we have the kinematic condition

$$\varphi_z^o - \varphi_x^o \zeta_x - \varphi_y^o \zeta_y = U\zeta_x + V^* \zeta_y \tag{8.59}$$

and the dynamic condition

$$-g\zeta = U\varphi_x^o + V^* \varphi_y^o + \frac{1}{2}(\varphi_x^{o2} + \varphi_y^{o2} + \varphi_z^{o2}). \tag{8.60}$$

Introducing the stretched variable $Z := z/\varepsilon$, we seek $\varphi^o(x, y, Z, \varepsilon)$ as a regular perturbation series in the form:

$$\varphi^o(x, y, Z, \varepsilon) = \varepsilon\varphi_1(x, y, Z) + \varepsilon^2\varphi_2(x, y, Z) + \cdots .$$

From the Laplace equation which becomes now in terms of Z,

$$\varphi_{ZZ}^o = -\varepsilon^2(\varphi_{xx}^o + \varphi_{yy}^o) =: -\varepsilon^2 \Delta_2 \varphi^o,$$

we find for φ_i,

$$\varphi_{i\,ZZ} = \begin{cases} 0, & i = 1, 2, \\ -\Delta_2 \varphi_{i-2}, & i = 3, 4, \ldots, \end{cases}$$

from which it follows that

$$\begin{aligned}
\varphi_1 &= \psi_1(x, y), \\
\varphi_2 &= \psi_2(x, y), \\
\varphi_3 &= \psi_3(x, y) - \frac{1}{2}(Z + h)^2 \Delta_2 \psi_1(x, y),
\end{aligned} \tag{8.61}$$

with $h = h^*/\varepsilon$. The ψ_i can be determined in the following way.

We begin with the dynamic free surface condition (8.60). This leads to the relation

$$-g\zeta = \varepsilon U \psi_{1x} + \varepsilon^2 (U \psi_{2x} + V \psi_{1y} + \psi_{1x}^2 + \psi_{1y}^2) + \mathrm{O}(\varepsilon^3), \tag{8.62}$$

with $V = V^*/\varepsilon$. Then (8.62) together with the fact that $F_{h^*}^2 = \mathrm{O}(1)$ implies that one may consider ζ in the form

$$\zeta = \varepsilon^2 \zeta_2 + \varepsilon^3 \zeta_3 + \cdots,$$

and obtain from (8.62) by equating like powers in ε,

$$\begin{aligned}
\zeta_2 &= -\frac{U}{\varepsilon g} \psi_{1x} = -\frac{\ell}{U} \psi_{1x}, \\
\zeta_2 &= -\frac{\ell}{U^2} \left\{ U \psi_{2x} + V \psi_{1y} + \frac{1}{2}(\psi_{1x}^2 + \psi_{1y}^2) \right\},
\end{aligned} \tag{8.63}$$

where ℓ is of O(1) and defined by $U^2/g = \varepsilon \ell$. Now it follows from the kinematic condition (8.59) together with (8.61) and (8.63) that

$$\begin{aligned}
&-\varepsilon^2 h \Delta_2 \psi_1 - \varepsilon^3 (\zeta_2 \Delta_2 \psi_1 + h \Delta_2 \psi_2) + \mathrm{O}(\varepsilon^4) \\
&\quad = \varepsilon^2 U \zeta_{2x} + \varepsilon^3 (U \zeta_{3x} + V \zeta_{2y} + \zeta_{2x} \psi_{1x} + \zeta_{2y} \psi_{1y}) + \mathrm{O}(\varepsilon^4).
\end{aligned}$$

Consequently, we obtain

$$\begin{aligned}
\Delta_2 \psi_1 &= -\frac{U}{h} \zeta_{2x}, \\
\Delta_2 \psi_2 &= -\frac{\zeta_2}{h} \Delta_2 \psi_1 - \frac{1}{h} \{ U \zeta_{3x} + V \zeta_{2y} + \zeta_{2x} \psi_{1x} + \zeta_{2y} \psi_{1y} \}.
\end{aligned} \tag{8.64}$$

Eliminating ζ_2 and ζ_3 from (8.64) and (8.63) yields:

$$\begin{aligned}
&(1 - F_{h^*}^2) \psi_{1xx} + \psi_{1yy} = 0, \quad \text{and} \\
&(1 - F_{h^*}^2) \psi_{2xx} + \psi_{2yy} = \Omega_2(x, y),
\end{aligned} \tag{8.65}$$

where $\Omega_2(x, y)$ is a function of ζ_2 and ψ_1 only. Equation (8.65) is the linearised shallow water equation which is a partial differential equation of mixed type depending on the value of F_{h^*}. We confine only to the elliptic case where $F_{h^*} < 1$.

Boundary conditions for ψ_1 follow from the matching procedure at $y = 0$ with the inner solution.

We now turn to the solution of (8.65) and write

$$\psi_1(x, y) = \psi_1^e(x, y) + \psi_1^o(x, y),$$

where ψ_1^e is an even function of y and ψ_1^o is odd in y. The even part may be written as a source distribution

$$\psi_1^e(x, y) = \frac{1}{2\pi\sqrt{1 - F_{h^*}^2}} \int_{-\infty}^{\infty} \sigma(\xi) \log\{(x - \xi)^2 + (1 - F_{h^*}^2)y^2\}^{1/2} \, \mathrm{d}\xi \tag{8.66}$$

and the odd part may be written as a vortex distribution

$$\psi_1^o(x, y) = \frac{1}{2\pi\sqrt{1 - F_{h^*}^2}} \int_{-\infty}^{\infty} \gamma(\xi) \arctan\left(\frac{y\sqrt{1 - F_{h^*}^2}}{x - \xi}\right) \mathrm{d}\xi. \tag{8.67}$$

Because of the proposed matching at $y = 0$, we must make a series expansion for small values of y for both ψ_1^e and ψ_1^o. We find the Taylor expansion series expansion

$$\psi_1^e(x, y) \approx \psi_1^e(x, 0) + |y|\frac{\sigma(x)}{2} - \frac{1}{2}y^2(1 - F_{h^*}^2)\psi_{1xx}^e(x, 0) + \cdots, \tag{8.68}$$

where

$$\psi_1^e(x, 0) = \frac{1}{2\pi\sqrt{1 - F_{h^*}^2}} \int_{-\infty}^{\infty} \sigma(\xi) \log|x - \xi| \, \mathrm{d}\xi.$$

For the odd part, we find

$$\psi_1^o(x, y) \approx \mathrm{sgn}(y)\Gamma(x) + \frac{y}{2\pi} \fint_{-\infty}^{\infty} \frac{\gamma(\xi)}{x - \xi} \, \mathrm{d}\xi + \mathrm{O}(y^2), \tag{8.69}$$

where $\Gamma(x) = -\frac{1}{2}\int_{-\infty}^{x} \gamma(\xi)\,\mathrm{d}\xi = -\frac{1}{2}\int_{-\infty}^{\infty} \gamma(\xi)\mathrm{H}(x - \xi)\,\mathrm{d}\xi$ with $\mathrm{H}(x - \xi)$ denoting the Heaviside step function.

Now we are in a position to consider the expansion of the total outer field $\Phi^o = \varphi_0^o + \varphi^o$ for small values of y, where φ_0^o and φ^o are the unperturbed and perturbed potentials as in (8.57). Collecting the results (8.57), (8.61), (8.68) and (8.69), we finally arrive at the expansion:

$$\begin{aligned}
\Phi^o &\approx Ux + \varepsilon\psi_1^e(x, 0) \pm \Gamma(x) \\
&\quad + y\left(V^* + \frac{\varepsilon}{2\pi} \fint_{-\infty}^{\infty} \frac{\gamma(\xi)\,\mathrm{d}\xi}{x - \xi}\right) \pm y\frac{\varepsilon\sigma(x)}{2} + \mathrm{O}(y^2) \\
&=: Ux + \varphi_1^e(x) \pm [\varphi_1^o](x) \\
&\quad + yV^o(x) \pm y[V^e](x) + \mathrm{O}(y^2), \quad \text{for } y \gtrless 0,
\end{aligned} \tag{8.70}$$

with $\varphi_1^e, [\varphi_1^o], V^o$ and $[V^e]$ defined accordingly. It should be emphasised that the densities σ and γ in the above formulation remain to be determined. For this purpose, we have to consider the problem in the inner region.

In the inner region we have

$$x = \mathrm{O}(1), \qquad y = \mathrm{O}(\varepsilon) \quad \text{and} \quad z = \mathrm{O}(\varepsilon).$$

Hence we introduce the stretched variables $Y = y/\varepsilon$ and $Z = z/\varepsilon$, in terms of which the Laplace equations now become

$$\varepsilon^2 \Phi^i_{xx} + \Phi^i_{YY} + \Phi^i_{ZZ} = 0,$$

where we have adapted the notion Φ^i for the inner potential $\Phi^i = \Phi^i(x, Y, Z, \varepsilon)$. We assume that Φ^i admits a regular expansion in the form

$$\Phi^i(x, Y, Z, \varepsilon) = \Phi_0(x, Y, Z) + \varepsilon \Phi_1(x, Y, Z) + \mathrm{O}(\varepsilon^2), \tag{8.71}$$

and substituting it into the Laplace equation, we obtain

$$\tilde{\Delta}_2 \Phi_i = \begin{cases} 0, & i = 1, 2, \\ -\Phi_{i-1xx} & i = 3, 4, \dots, \end{cases}$$

where $\tilde{\Delta}_2 := \partial^2/\partial Y^2 + \partial^2/\partial Z^2$. The free surface conditions (8.59) and (8.60) yield for the first approximations

$$\Phi_{0Z} = 0, \qquad \Phi_{1Z} = 0, \quad \text{at } Z = 0.$$

Again we split up the inner potential Φ^i into Φ^{ie} and Φ^{io}, the even and the odd part of Φ^i. The even potential relates to the thickness problem; since the even part of the unperturbed flow equals Ux, we take $\Phi^e_0 = Ux$. Furthermore, Φ^{ie} has to fulfil the boundary condition on the ship hull, i.e.,

$$\frac{\partial \Phi^{ie}}{\partial n} = 0, \quad \text{on } y = \pm f(x, z) = \pm \varepsilon F(x, Z). \tag{8.72}$$

We now define $\mathbf{N}^*$ to be the projection of the normal $\mathbf{n}$ on the plane $x =$ constant (see Fig. 8.5).

From (8.71) for $y > 0$, we have

$$\Phi^e_x f_x + \Phi^e_z f_z - \Phi^e_y = 0.$$

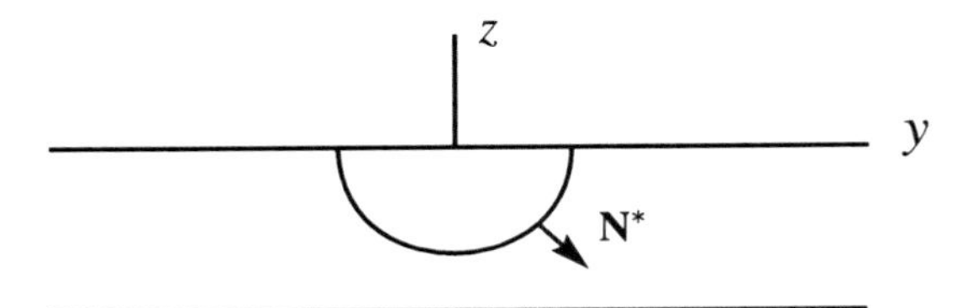

Fig. 8.5 Local configuration

We also have

$$\frac{\partial \Phi^{ie}}{\partial N^*} = \frac{\Phi^{ie}_y - \Phi^{ie}_z f_z}{\sqrt{1+f_z^2}},$$

from which it follows that

$$\frac{\partial \Phi^{ie}}{\partial N^*} = \frac{\Phi^{ie}_x f_x}{\sqrt{1+f_z^2}}. \tag{8.73}$$

Inserting (8.71) into (8.73) and defining the normal **N** in terms of the stretched coordinates $Y, Z, \frac{\partial}{\partial N^*} = \frac{1}{\varepsilon}\frac{\partial}{\partial N}$, we then find with $f = \varepsilon F$ that

$$\frac{\partial \Phi^e_1}{\partial N} = 0,$$
$$\frac{\partial \Phi^e_1}{\partial N} = \frac{UF_x}{\sqrt{1+F_z}} =: V_N.$$

Therefore, Φ^e_1 and Φ^e_2 take the forms

$$\begin{aligned} \Phi^e_1 &= g_1(x), \\ \Phi^e_2 &= g_2(x) + \Phi^*_2(x, Y, Z). \end{aligned} \tag{8.74}$$

This inner solution has to be matched with the outer one. Therefore Φ^*_2 must be known for large values of Y and there is no need to solve the equation for Φ^*_2 . For this purpose, we assume Φ^*_2 has the asymptotic behaviour

$$\Phi^*_2 \sim V|Y|, \quad \text{as } Y \to \infty,$$

where V may be calculated simply by conservation of mass. We proceed as follows.

As indicated in Fig. 8.6, the apparent flux out of the ship is given by

$$\int V_N \, \mathrm{d}l = US'(x),$$

where S' is the derivative of the surface area of the stretched cross section. Hence

$$V = \frac{U}{2h} S'(x), \quad \text{for } |x| < L/2.$$

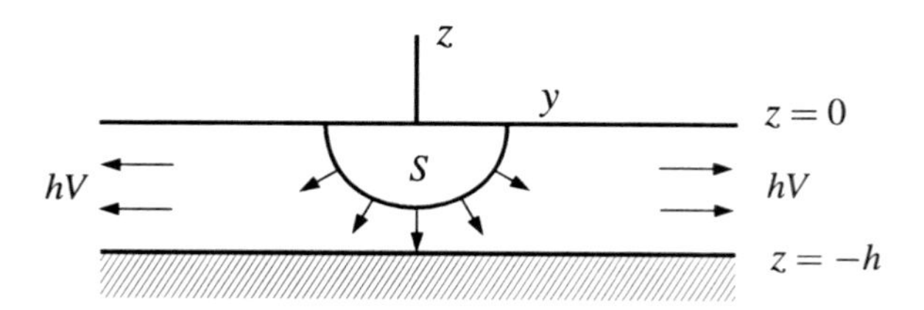

Fig. 8.6 Cross flow

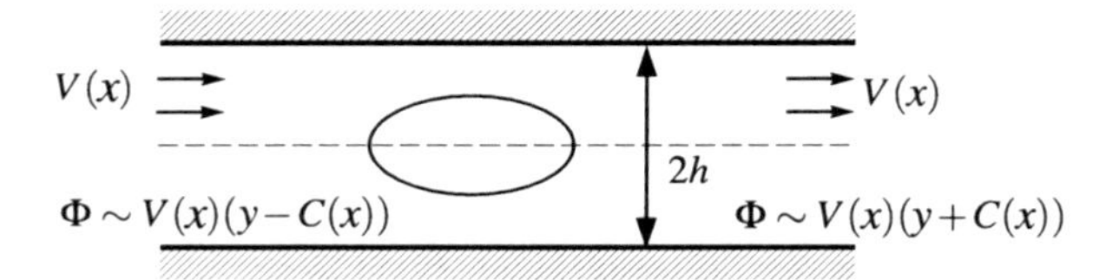

Fig. 8.7 Local flow around the double body

We now match the even part of the inner and outer solution (8.68), (8.74), and find

$$\sigma(x) = \frac{U}{h} S'(x) \tag{8.75}$$

and

$$g_1(x) = \psi_1^e(x, 0) = \frac{U}{2\pi h\sqrt{1 - F_{h^*}^2}} \int_{-L/2}^{L/2} S'(\xi) \log|x - \xi| \, d\xi. \tag{8.76}$$

The odd part of the solution relates to the cross flow $V^o(x)y$ given by (8.70). The inner problem we have to consider is the same as in Sect. 8.2, if we reflect the physical domain with respect to $y = 0$ (see Fig. 8.7).

In this figure, $C(x)$ is the known blockage function or coefficient as introduced in (8.45) and $V(x)$ is to be determined by the matching procedure below. The far-field behaviour of the inner solution can be matched with the local behaviour of the outer solution. We obtain, from (8.70),

$$[\Phi_1^o](x) = V(x)C(x), \quad \text{and}$$

$$V^o(x) = V(x).$$

Consequently, we have

$$-\frac{\varepsilon}{2} \int_{-\infty}^{\infty} \gamma(\xi) H(x - \xi) \, d\xi = V(x)C(x) \tag{8.77}$$

and

$$V^* + \frac{\varepsilon}{2\pi} ⨍_{-\infty}^{\infty} \frac{\gamma(\xi)}{x - \xi} \, d\xi = V(x), \tag{8.78}$$

from which it follows that $\gamma(\xi)$ has to be a solution of the following singular integral equation:

$$\frac{\varepsilon}{2\pi} ⨍_{-\infty}^{\infty} \left\{ \frac{1}{x - \xi} + \pi \frac{H(x - \xi)}{C(x)} \right\} \gamma(\xi) \, d\xi = -V^* := -\varepsilon V. \tag{8.79}$$

We can easily obtain an integral equation for $V(x)$, because differentiation of (8.77) with respect to x leads to

$$\varepsilon\gamma(x) = -2(VC)'.$$

Since in this section we consider the case of the steady motion, it follows that $\gamma(x) = 0$, for $|x| > L/2$.

We shall show that (8.79) can be transposed into a Fredholm integral equation of the second kind, which can be solved numerically.

Consider

$$\frac{1}{\pi \mathrm{i}} \int_{-L/2}^{L/2} \frac{\gamma(\xi)}{x-\xi} \, \mathrm{d}\xi = 2\mathrm{i}V + \frac{1}{C(x)} \int_{-L/2}^{x} \gamma(\xi) \, \mathrm{d}\xi := \mathrm{i}F_\gamma(x). \tag{8.80}$$

We now use the theory of Hilbert transforms (see Sect. 9.3) again and consider

$$\frac{1}{\pi \mathrm{i}} \int_{-L/2}^{L/2} \frac{\gamma(\xi)}{x-\xi} \, \mathrm{d}\xi = \mathrm{i}F_\gamma(x).$$

We apply the Kutta condition at $x = L/2$ to obtain

$$\gamma(x) = \frac{1}{\pi} \sqrt{\frac{x - L/2}{x + L/2}} \int_{-L/2}^{L/2} \sqrt{\frac{\zeta + L/2}{\zeta - L/2}} \frac{F_\gamma(\zeta)}{x - \zeta} \, \mathrm{d}\zeta. \tag{8.81}$$

Equation (8.81) now is a Fredholm integral equation of the second kind for $\gamma(x)$. It can be shown that the kernel is square integrable and hence the equation has a unique solution. This unique solution can be found numerically. In particular, if $C(x) \gg 1$, one can define a sequence $\{\gamma_i\}$ by the iteration procedure

$$\gamma_i(x) = \frac{1}{\pi} \sqrt{\frac{x - L/2}{x + L/2}} \int_{-L/2}^{L/2} \sqrt{\frac{\zeta + L/2}{\zeta - L/2}} \frac{1}{x - \zeta} \left[2V + \frac{1}{C(\zeta)} \int_{-L/2}^{\zeta} \gamma_{i-1}(\xi) \, \mathrm{d}\xi \right] \mathrm{d}\zeta. \tag{8.82}$$

This sequence can be used to approximate the exact solution.

In the case of total blockage $C(x) = \infty$ we have obtained the same result as in the 2-D airfoil theory for steady flow. Equation (8.82) with $C = \infty$ is identical to (8.41) if $P(x)$ is determined so that the Kutta condition $\gamma(L/2) = 0$ is fulfilled. It will be clear that the theory of this section may be extended to non-steady flow problems in a similar way as in Sect. 8.1. We then have to use the same conditions to obtain a unique vortex distribution.

To obtain the heave force and pitch moment we shall integrate the local pressure distribution along the hull. The local field up to first order in ε is described by

$$\Phi^i \approx Ux + \varepsilon \psi_1^e(x, 0) \pm \Gamma(x) \tag{8.83}$$

and for the determination of the heave and pitch, we may disregard the term $\pm\Gamma(x)$. We use Bernoulli's equation and find that the even part of the first-order pressure along the ship is given by

$$p^e \sim -\rho U \varepsilon \psi_{1x}^e(x, 0). \tag{8.84}$$

The heave force becomes

$$F_3 = -\rho U \varepsilon \int_{-L/2}^{L/2} \psi_{1x}^e B(x)\,\mathrm{d}x, \tag{8.85}$$

and the pitch moment

$$F_5 = -\rho U \varepsilon \int_{-L/2}^{L/2} \psi_{1x}^e B(x) x\,\mathrm{d}x, \tag{8.86}$$

where $B(x)$ is the beam of the ship at the location x.

The sway force and yaw moment can be calculated similarly. From (8.83), the contribution from the odd part $\pm\varepsilon\Gamma(x)$ leads to

$$p^o(x) \sim \pm\rho U \varepsilon \Gamma_x(x).$$

Hence the sway force (lift) is given by

$$F_2 = -\rho U h \varepsilon \int_{-L/2}^{L/2} \gamma(x)\,\mathrm{d}x. \tag{8.87}$$

The yaw moment now becomes

$$F_6 = -\rho U h \varepsilon \int_{-L/2}^{L/2} \gamma(x) x\,\mathrm{d}x. \tag{8.88}$$

The sway force and yaw moment may also be calculated by means of the Blasius formulas. In that case, we write the outer potential as the real part of the complex potential

$$\chi(z) = \Phi(z) + \mathrm{i}\Psi(z), \quad \text{where } z = x + \mathrm{i}y.$$

The forces acting on the ship are then

$$F_1 - \mathrm{i}F_2 = \frac{\mathrm{i}\rho}{2} \oint_{\mathscr{C}} \left(\frac{\mathrm{d}W}{\mathrm{d}z}\right)^2 \mathrm{d}z, \tag{8.89}$$

and the moment

$$F_6 = -\Re\left\{\frac{\rho}{2} \oint_{\mathscr{C}} z \left(\frac{\mathrm{d}W}{\mathrm{d}z}\right)^2 \mathrm{d}z\right\}, \tag{8.90}$$

where $\mathscr{C}$ is a closed contour around the object. In our case we considered

$$W(z) = (U + \mathrm{i}V^*)z + \text{source dist.} + \text{vortex dist.},$$

and if we equate the contributions of the residues, it follows that $F_1 = 0$ while F_2 and F_6 have the forms as given by (8.87) and (8.88), respectively. However we would like to mention that this derivation only holds for steady flow fields.

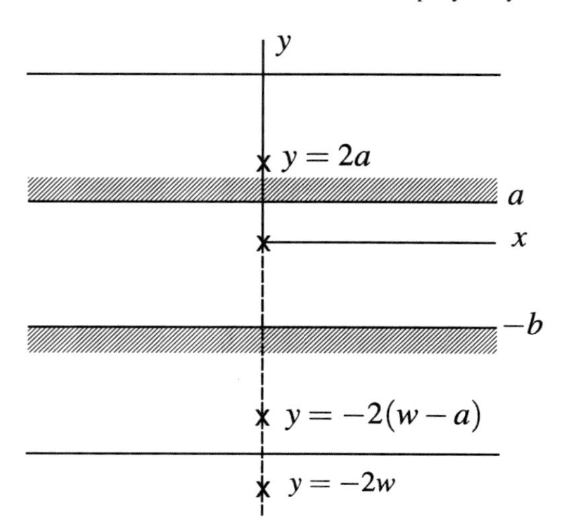

Fig. 8.8 Reflection of singularities

8.4 Ships in a Channel

In this section we will show how we can apply the theory developed in the previous section to the situation of a ship in a channel. Since most of the analysis we considered here is exactly the same as before, we omit the details. We consider the case that the side walls of the channel are in the outer region. The outer field may be considered as a superposition of sources and vortices. First we shall give a derivation of the potential due to a source of unit strength situated between two parallel walls.

We introduce the complex variable

$$x = x + \mathrm{i}\beta y, \quad \text{with } \beta = \sqrt{1 - F_h^2},$$

where F_h is the Froude number with respect to the depth h of the channel. Throughout the section the side walls of the channel are at $y = a$ and $y = -b$. The total width of the channel is denoted by w as is shown in Fig. 8.8.

We consider a source with unit strength located at the origin. The field due to this source may be obtained by means of reflection to the walls. In this case, the complex potential denoted by $\chi(z)$ has the form

$$\begin{aligned}\chi(z) &= \Phi + \mathrm{i}\Psi \\ &= \frac{1}{2\pi}\log z + \frac{1}{2\pi}\sum_{n=1}^{\infty}\log(z - 2\beta\mathrm{i}nw) + \frac{1}{2\pi}\sum_{n=1}^{\infty}\log(z + 2\beta\mathrm{i}nw) \\ &\quad + \frac{1}{2\pi}\log(z - 2\beta\mathrm{i}a) + \frac{1}{2\pi}\sum_{n=1}^{\infty}\log(z - 2\beta\mathrm{i}(nw + a)) \\ &\quad + \frac{1}{2\pi}\sum_{n=1}^{\infty}\log(z + 2\beta\mathrm{i}(nw - a)).\end{aligned} \tag{8.91}$$

We now consider the first three terms in (8.91) and write

$$\chi_1 = \frac{1}{2\pi}\log z + \frac{1}{2\pi}\sum_{n=1}^{\infty}\log(z - 2\beta \mathrm{i} n w) + \frac{1}{2\pi}\sum_{n=1}^{\infty}\log(z + 2\beta \mathrm{i} n w)$$
$$= \frac{1}{2\pi}\log\left[z\prod_{n=1}^{\infty}\left(1 - \frac{z^2}{(2\mathrm{i}\beta n w)^2}\right)\cdot \text{constant}\right].$$

The function

$$z\prod_{n=1}^{\infty}\left(1 - \frac{z^2}{(2\mathrm{i}\beta n w)^2}\right)$$

is an entire function with simple zeros at $z = \pm 2\mathrm{i}nw$. The lemma of Weierstrass allows us to write

$$z\prod_{n=1}^{\infty}\left(1 - \frac{z^2}{(2\mathrm{i}\beta n w)^2}\right) = \frac{2\mathrm{i}\beta w}{\pi}\sin\frac{z\pi}{2\mathrm{i}\beta w} = \frac{2\beta w}{\pi}\sinh\frac{z\pi}{2\beta w}.$$

Hence up to an additive constant we can simplify χ_1 so that

$$\chi_1 = \frac{1}{2\pi}\log[\sinh k_0 z],$$

with $k_0 = \frac{\pi}{2\beta w}$, $z = x + \mathrm{i}\beta y$ and $\beta = \sqrt{1 - F_h^2}$. The remaining three terms in (8.91) lead to a similar result, and the complete potential due to a unit source at $z = \zeta$ becomes

$$G_s(z;\zeta) = \frac{1}{2\pi}\left\{\log[\sinh(k_0(z-\zeta))] + \log[\sinh(k_0(z-\zeta) - \mathrm{i}\alpha)]\right\} \tag{8.92}$$

with $\alpha = \frac{\pi a}{w}$. The complex potential due to a unit vortex at the point $z = \zeta$ becomes

$$G_v(z;\zeta) = -\frac{\mathrm{i}}{2\pi}\left\{\log[\sinh(k_0(z-\zeta))] - \log[\sinh(k_0(z-\zeta) - \mathrm{i}\alpha)]\right\}. \tag{8.93}$$

The perturbation potential in the outer region (see (8.57)) can now be written as a distribution of sources and vortices along the axis of the ship:

$$\varphi^o(x,y) = \Re\int_{-L/2}^{L/2}\left[\sigma(\xi)G_s(z;\xi) + \gamma(\xi)G_v(z;\xi)\right]\mathrm{d}\xi, \tag{8.94}$$

where ξ is the coordinate along the axis of the ship. Here again σ and γ are the unknown densities as in (8.66) and (8.67). They will be determined by the matching conditions from the inner potential Φ^i as in Sect. 8.3. To do so we introduce, for convenience, the (s,t) coordinates fixed to the ship and let (U, V) be the components of the uniform flow $\mathbf{U}$ with respect to the coordinates (s,t), as shown in

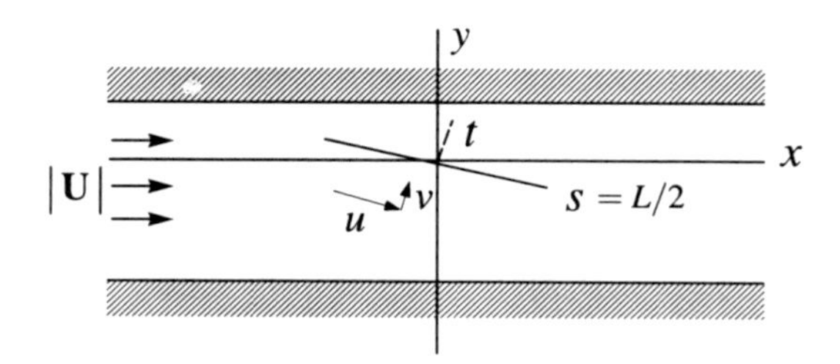

Fig. 8.9 Outer coordinates fixed to ship

Fig. 8.9. Furthermore we assume the angle of attack to be small and expand (8.94) for small t.

This leads after some lengthy calculations to

$$\varphi^i(s,t) \sim Us + Vt + \varphi_1(s) \pm [\varphi_1](s) + t(V(s) \pm [V](s)), \tag{8.95}$$

where

$$\begin{aligned}
\varphi_1(s) = \frac{1}{2\pi}\int_{-L/2}^{L/2} \mathrm{d}\xi \Bigg(\sigma(\xi)\Bigg\{ &\log[\sinh(k_0(s-\xi))] \\
&+ \frac{1}{2}\log[\sinh(k_0(s-\xi) + \sin^2\alpha)]\Bigg\} \\
&+ \gamma(\xi)\arctan\left[\frac{\cosh(k_0(s-\xi))\sin\alpha}{\sinh(k_0(s-\xi))\cos\alpha}\right]\Bigg), \\
[\varphi_1] = -\frac{1}{2}\int_{-L/2}^{L/2} &\gamma(\xi)H(s-\xi)\,\mathrm{d}\xi, \\
V(s) = \frac{1}{4w}\int_{-L/2}^{L/2} \mathrm{d}\xi \Bigg\{ &\frac{-\sigma(\xi)\sin 2\alpha}{\cosh(2k_0(s-\xi)) - \cos 2\alpha} \\
&+ \gamma(\xi)\frac{\cosh(k_0(s-\xi))(1-\cos 2\alpha)}{\cosh(2k_0(s-\xi)) - \cos 2\alpha}\Bigg\}, \\
[V(s)] = \beta\frac{\sigma(s)}{2}&,
\end{aligned}$$

and $H(s-\xi)$ is the Heaviside step function (cf. (8.70).

By following the arguments used in the previous section, the solution in the inner region can be written in the form

$$\varphi^i \sim Us + F_1(s) \pm \frac{US'(s)}{2h}t + V_1(s)[t \pm C(s)], \tag{8.96}$$

where F_1 and V_1 are to be determined. Here S is the surface area and C the blockage coefficient as before. This expression must be equal to (8.95) which leads to

$$\begin{aligned}
&\text{(a)} \quad \varphi_1(s) = F_1(s),\\
&\text{(b)} \quad [\varphi_1](s) = V_1(s)C(s),\\
&\text{(c)} \quad V + V(s) = V_1(s),\\
&\text{(d)} \quad [V](s) = \frac{US'(s)}{h}.
\end{aligned}$$

The last condition gives us the source strength in (8.94),

$$\sigma(s) = \frac{US'(s)}{h}.$$

It remains now to determine $\gamma(\xi)$ from the other relations. Conditions (b) and (c) imply

$$\begin{aligned}
&⨍_{-L/2}^{L/2} \gamma(\xi) \left\{ \frac{1-\cos 2\alpha}{\cosh(2k_0(s-\xi)) - \cos 2\alpha} \coth(k_0(s-\xi)) + \frac{2w}{C(s)} H(s-\xi) \right\} \mathrm{d}\xi \\
&= -\frac{\pi U}{wh\beta^2} \sin 2\alpha \int_{-L/2}^{L/2} \frac{S(\xi)\sinh(2k_0(s-\xi))}{[\cosh(2k_0(s-\xi)) - \cos 2\alpha]^2} \mathrm{d}\xi - 4wV := f(s).
\end{aligned}$$

This is an integral equation of the Cauchy type, therefore we write it in the following standard form (see Sect. 9.3):

$$\frac{1}{\pi \mathrm{i}} \int_{-L/2}^{L/2} \frac{\gamma(\xi)}{s-\xi} \mathrm{d}\xi = F_\gamma(s), \tag{8.97}$$

where

$$\begin{aligned}
F_\gamma(s) := \frac{k_0}{\pi \mathrm{i}} f(s) - \frac{2wk_0}{\pi \mathrm{i} C(s)} \int_{-L/2}^{s} \gamma(\xi)\,\mathrm{d}\xi \\
- \frac{1}{\pi \mathrm{i}} \int_{-L/2}^{L/2} \gamma(\xi) \frac{A(s,\xi)k_0(s-\xi)\cosh(k_0(s-\xi)) - 1}{s-\xi} \mathrm{d}\xi
\end{aligned}$$

with

$$A(s,\xi) := \frac{1-\cos 2\alpha}{\cosh(k_0(s-\xi) - \cos 2\alpha}.$$

From (8.97) we obtain a Fredholm integral equation of the second kind. The unique solution which obeys the Kutta condition at $s = L/2$ takes the form

$$\gamma(s) = \frac{1}{\pi \mathrm{i}} \sqrt{\frac{s-L/2}{s+L/2}} ⨍_{-L/2}^{L/2} \sqrt{\frac{\zeta + L/2}{\zeta - L/2}} \frac{F_\gamma(\zeta)}{s-\zeta} \mathrm{d}\zeta.$$

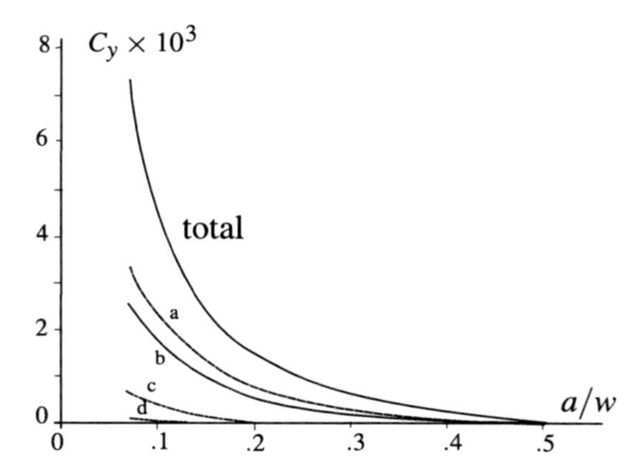

Fig. 8.10 (**a**) first-order, (**b**) 2nd-order source term, (**c**) 2nd-order source-vortex term, (**d**) 2nd-order vortex term. See Beck [1]

By a similar argument as in Sect. 8.3, the sway force and yaw moment can be calculated:

$$
\begin{aligned}
F_2 = & -\rho U h \int_{-L/2}^{L/2} \gamma(\xi)\, \mathrm{d}\xi \\
& + \frac{\rho U}{2w\beta^2} \int_{-L/2}^{L/2} \int_{-L/2}^{L/2} S'(\xi)\gamma(\zeta) \frac{\sinh(2k_0(\xi-\zeta))}{\cosh(2k_0(\xi-\zeta)) - \cos 2\alpha}\, \mathrm{d}\xi\, \mathrm{d}\zeta \\
& + \frac{\rho U^2}{4w\beta^3 h} \sin 2\alpha \int_{-L/2}^{L/2} \int_{-L/2}^{L/2} S'(\xi) S'(\zeta) \frac{1}{\cosh(2k_0(\xi-\zeta)) - \cos 2\alpha}\, \mathrm{d}\xi\, \mathrm{d}\zeta \\
& + \frac{\rho h}{4w\beta} \sin 2\alpha \int_{-L/2}^{L/2} \int_{-L/2}^{L/2} \gamma(\xi)\gamma(\zeta) \frac{1}{\cosh(2k_0(\xi-\zeta)) - \cos 2\alpha}\, \mathrm{d}\xi\, \mathrm{d}\zeta, \\
F_6 = & -\rho U h \int_{-L/2}^{L/2} \xi\gamma(\xi)\, \mathrm{d}\xi - \frac{\rho U}{2w\beta^2} \int_{-L/2}^{L/2} \int_{-L/2}^{L/2} S'(\xi)\gamma(\zeta) \\
& \cdot \left\{ \frac{(\xi-\zeta)\sin^2\alpha \coth(k_0(\xi-\zeta)) - \zeta \sinh(2k_0(\xi-\zeta))}{\cosh(2k_0(\xi-\zeta)) - \cos 2\alpha} \right\} \mathrm{d}\xi\, \mathrm{d}\zeta \\
& + \frac{\rho U^2}{8w\beta^3 h} \sin 2\alpha \int_{-L/2}^{L/2} \int_{-L/2}^{L/2} S'(\xi) S'(\zeta) \frac{\xi+\zeta}{\cosh(2k_0(\xi-\zeta)) - \cos 2\alpha}\, \mathrm{d}\xi\, \mathrm{d}\zeta \\
& + \frac{\rho h}{8w\beta} \sin 2\alpha \int_{-L/2}^{L/2} \int_{-L/2}^{L/2} \gamma(\xi)\gamma(\zeta) \frac{\xi+\zeta}{\cosh(2k_0(\xi-\zeta)) - \cos 2\alpha}\, \mathrm{d}\xi\, \mathrm{d}\zeta.
\end{aligned}
$$

To illustrate the idea, an example of calculations, carried out by Beck [1], based on these formulas is given in Fig. 8.10. It turns out that the interaction terms between source and vortex distribution of the reflected points lead to large contributions in the sway force. These terms are asymptotically of lower order. However, due to summation, they become numerically of the same order of magnitude.

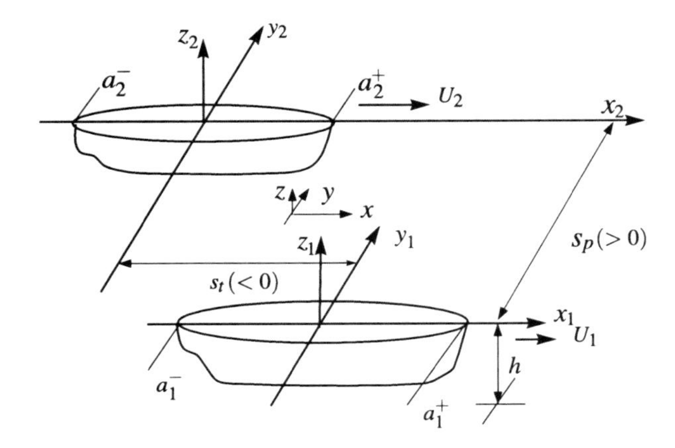

Fig. 8.11 Passing ships

8.5 Interaction of Ships

To conclude this chapter, we now consider the interaction of ships in shallow water. To simplify the presentation, we describe the situation as indicated in Fig. 8.11. In particular we assume that the lateral distance $s_p = \mathrm{O}(1)$ as $\varepsilon \to 0$, where ε is the slenderness parameter. As in Fig. 8.11, ship 1 travels at speed U_1 while the speed of the second ship is U_2. We consider a speed range for both ships such that $F_h^2 = U^2/gh = \mathrm{O}(\varepsilon)$. Hence, the outer potentials of both ships satisfy the Laplace equation with respect to a fixed coordinate system:

$$\varphi_{xx}^o + \varphi_{yy}^o = 0.$$

The outer solution may be considered as a source distribution along the centre line of the ships together with a vortex distribution along the centre line and the wakes of the ships. The source and vortex strengths are time dependent. We have

$$\varphi^o(x,y,t) = \sum_{j=1}^{2} \left\{ \frac{1}{4\pi} \int_{a_j^-(t)}^{a_j^+(t)} \sigma_j(\xi,t) \log\left[(x-\xi)^2 + \left(y - (-1)^j \frac{s_p}{2} \right)^2 \right] \mathrm{d}\xi \right.$$
$$\left. + \frac{1}{2\pi} \int_{-\infty}^{a_j^+(t)} \gamma_j(\xi,t) \arctan\left[\frac{y-(-1)^j \frac{s_p}{2}}{x-\xi} \right] \mathrm{d}\xi \right\}. \tag{8.98}$$

We know that in the wake the vortices are transported along the stream line with the local velocity as in (8.5). This means

$$\frac{\partial \gamma_j}{\partial t}(x,t) = 0. \tag{8.99}$$

Furthermore, it follows from Kelvin's theorem that

$$\frac{\mathrm{d}\Gamma_j}{\mathrm{d}t} = -U_j \gamma_j(a_j^-(t)), \tag{8.100}$$

where Γ_j is the total circulation around the ship. To obtain uniqueness we require the Kutta condition:

$$\lim_{x_j \to a_j^- - 0} \gamma_j = \lim_{x_j \to a_j^- + 0} \gamma_j \tag{8.101}$$

holds for all t. Now we consider the field near ship 1 where (8.98) may be expanded in a Taylor series with respect to the ship coordinates (x_1, y_1):

$$\begin{aligned} \varphi^o(x_1, y_1, t) = {} & \varphi_1(x_1, \pm 0, t) \pm [\varphi_1](x, t) + V_{21}(x_1) y \pm \frac{\sigma_1(x_1, t)}{2} |y_1| \\ & + \frac{y_1}{2\pi} \int_{-\infty}^{a_1^+} \frac{\gamma_1}{x - \xi} \, \mathrm{d}\xi + \mathrm{O}(y_1^2), \quad \text{for } a_1^- \le x_1 \le a_1^+, \end{aligned} \tag{8.102}$$

where

$$\begin{aligned} \varphi_1(x, \pm 0, t) = {} & \frac{1}{2\pi} \int_{a_1^-(t)}^{a_1^+(t)} \sigma_1(\xi, t) \log |x_1 - \xi| \, \mathrm{d}\xi \\ & + \frac{1}{4\pi} \int_{a_2^-(t)}^{a_2^+(t)} \sigma_2(\xi, t) \log[(x_2 - \xi)^2 + s_p^2] \, \mathrm{d}\xi \\ & + \frac{1}{2\pi} \int_{-\infty}^{a_2^+(t)} \gamma_2(\xi, t) \arctan\left(\frac{s_p}{x_2 - \xi} \right) \mathrm{d}\xi, \\ [\varphi_1](x_1, t) = {} & \frac{1}{2} \int_{x_1}^{a_1^+(t)} \gamma_1(\xi, t) \, \mathrm{d}\xi, \end{aligned} \tag{8.103}$$

and the induced normal velocity

$$\begin{aligned} V_{21}(x_1, t) = {} & -\frac{1}{2\pi} \int_{a_2^-(t)}^{a_2^+(t)} \sigma_2(\xi, t) \frac{s_p}{(x_2 - \xi)^2 + s_p^2} \, \mathrm{d}\xi \\ & + \frac{1}{2\pi} \int_{-\infty}^{a_2^+(t)} \gamma_2(\xi, t) \frac{x_2 - \xi}{(x_2 - \xi)^2 + s_p^2} \, \mathrm{d}\xi. \end{aligned} \tag{8.104}$$

The solution in the inner region near ship 1 or 2 now again has an outer expansion

$$\varphi_j^i(x_j, y_j, z_j, t) = -\frac{U_j S_j'(x_j)}{4h} |y_j| + V_j^*(x_j, t)[y_j \pm C_j(x_j)] + f_j(x_j, t).$$

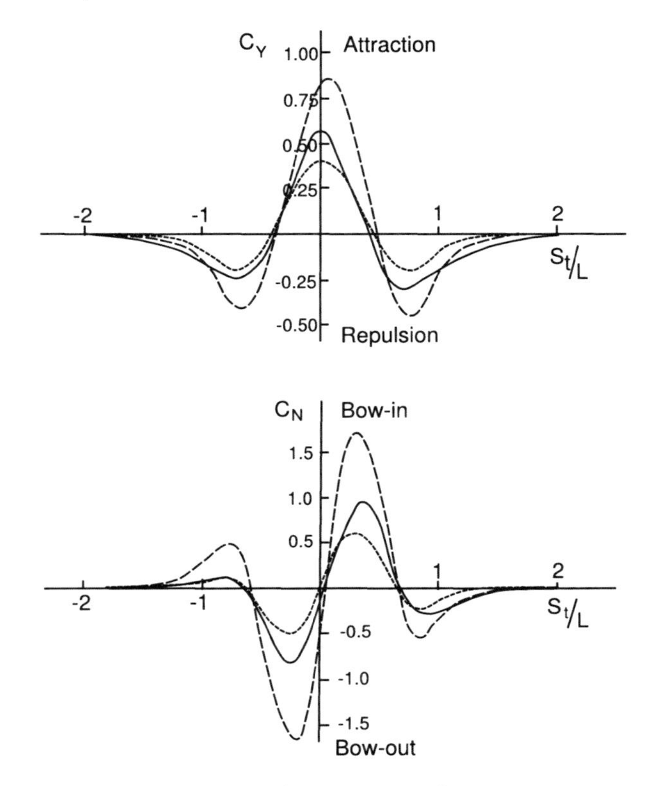

Fig. 8.12 —— experiments, – – – – and ······ computations

Matching the inner and the outer solution of ship 1 now leads to

$$
\begin{aligned}
\sigma_1(x_1, t) &= -\frac{U_1 S_1'(x_1)}{2h}, \\
V_1^*(x_1, t) &= V_{21}(x_1, t) + \frac{1}{2\pi} \int_{-\infty}^{a_1^+} \frac{\gamma_1(\xi, t)}{x_1 - \xi} \, \mathrm{d}\xi, \\
f_1(x_1, t) &= \varphi_1(x_1, \pm 0, t), \\
V_1^*(x_1, t) C_1(x_1, t) &= \frac{1}{2} \int_{x_1}^{a_1^+} \gamma_1(\xi, t) \, \mathrm{d}\xi.
\end{aligned}
\tag{8.105}
$$

We obtain an integral equation for $\gamma_1(\xi, t)$ similar to (8.70):

$$
\frac{1}{2\pi} \int_{-\infty}^{a_1^+} \left[\frac{\gamma_1(\xi, t)}{x_1 - \xi} - \frac{\pi \gamma_1(\xi, t)}{C_1(x_1)} H(\xi - x_1) \right] \mathrm{d}\xi = -V_{21}(x_1, t), \quad \text{for } a_1^- \leq x_1 \leq a_1^+. \tag{8.106}
$$

In a similar way we find near ship 2,

$$\frac{1}{2\pi}\int_{-\infty}^{a_2^+}\left[\frac{\gamma_2(\xi,t)}{x_2-\xi}-\frac{\pi\gamma_2(\xi,t)}{C_2(x_2)}H(\xi-x_2)\right]\mathrm{d}\xi=-V_{12}(x_2,t),$$

$$\text{for } a_2^- \le x_2 \le a_2^+. \tag{8.107}$$

Equations (8.106) and (8.107) are coupled singular integral equations for γ_1 and γ_2. To solve these equations, we may use a procedure with constant time step Δt. We start at a point $t = 0$ where no interaction takes place. This means that $\gamma_j = 0$ in the wake. We then start by taking V_{21} and V_{12} as the induced velocities due to the source distributions only. The integral equations may then be reduced to Fredholm equations with square integrable kernel (see Sect. 8.3) and can be solved numerically. The next time step gives us a constant vortex distribution along a wake of length $\Delta t U_j$ and the procedure can be repeated successively.

The sway force and yawing moment can be calculated by means of (8.43) and (8.44). An example of computations by Yeung [23] and experiments of Remery are shown in Fig. 8.12. The computations are done by means of two different approximations of the solution of the integral equations.

Chapter 9
Appendices: Mathematical Methods

In this chapter we present the derivations of some mathematical tools used in this book. No proofs of validity are given. We have used the method of stationary phase and the method of characteristics singular integral equations, for instance. Here we give a short introduction to these methods. We also give the derivation of a two-dimensional Green's function and a simplification of a set of algebraic equations used in Chap. 6.

9.1 The Method of Stationary Phase

This method dates back to Lord Kelvin, who developed it in particular for the case which is considered here. Suppose we have to evaluate an integral of the form

$$I = \int_{\alpha}^{\beta} \mathrm{e}^{\mathrm{i}\omega f(x)} g(x)\,\mathrm{d}x,$$

where $f(x)$ and $g(x)$ are differentiable real functions and ω is a large parameter. Integrating by parts, we find, if $f'(x) \neq 0$ in the interval $\alpha \leq x \leq \beta$,

$$\begin{aligned} I &= \frac{1}{\mathrm{i}\omega} \int_{\alpha}^{\beta} \frac{g(x)}{f'(x)} \,\mathrm{d}\big(\mathrm{e}^{\mathrm{i}\omega f(x)}\big) \\ &= \frac{1}{\mathrm{i}\omega} \left[\frac{g(x)}{f'(x)} \big(\mathrm{e}^{\mathrm{i}\omega f(x)}\big) \right]_{\alpha}^{\beta} - \frac{1}{\mathrm{i}\omega} \int_{\alpha}^{\beta} \mathrm{e}^{\mathrm{i}\omega f(x)} \frac{\mathrm{d}}{\mathrm{d}x} \left(\frac{g(x)}{f'(x)} \right) \mathrm{d}x. \end{aligned}$$

Repeating this procedure as many times as differentiations permit, we have a series in inverse powers of ω, which gives an asymptotic expansion for I. Obviously, this method breaks down if a point γ in the interval $\alpha \leq x \leq \beta$, $f'(\gamma) = 0$. The point γ is usually referred to as the stationary point. For this case the method of stationary phase is developed.

A.J. Hermans, *Water Waves and Ship Hydrodynamics*,
DOI 10.1007/978-94-007-0096-3_9,

Without loss of generality, let us consider an integral of the form

$$J(0, x_0) = \int_0^{x_0} \mathrm{e}^{\mathrm{i}\omega f(x)} g(x)\,\mathrm{d}x, \quad x_0 > 0, \tag{9.1}$$

where $x = 0$ is the only stationary point of f. Here f and g have as many derivatives as we need, and again ω is a large parameter. At the origin $f(x)$ now has an expansion

$$f(x) = f(0) + \frac{1}{2!}x^2 f''(0) + \frac{1}{3!}f'''(0) + \cdots.$$

We assume that $f''(0) > 0$; if $f''(0) < 0$ the argument has to be slightly changed. We introduce a new variable z defined by

$$z = f(x) - f(0),$$

and consider x as a function of z. From the fact that $f'(0) = 0$, we see that x admits an expansion of the form

$$x = z^{1/2}\sqrt{\frac{2}{f''(0)}}\{1 + C_1\sqrt{z} + C_2(\sqrt{z})^2 + \cdots\}.$$

In terms of the new variable z, the integral J takes the form

$$\mathrm{e}^{\mathrm{i}\omega f(0)} \int_0^{f(x_0)-f(0)} \mathrm{e}^{\mathrm{i}\omega z}\left(g(x)\frac{\mathrm{d}x}{\mathrm{d}z}\right)\mathrm{d}z. \tag{9.2}$$

Now if $g(x)\frac{\mathrm{d}x}{\mathrm{d}z}$ is expanded into powers of $\sqrt{z}$, we obtain

$$g(x)\frac{\mathrm{d}x}{\mathrm{d}z} = \frac{g(0)}{\sqrt{2f''(0)}} z^{-1/2}\{1 + \alpha_1\sqrt{z} + \alpha_2(\sqrt{z})^2 + \cdots\}, \tag{9.3}$$

which shows that we have to evaluate integrals of the form

$$K_n(a) := \int_0^a \mathrm{e}^{\mathrm{i}\omega z} z^{(n-1)/2}\,\mathrm{d}z, \quad a = f(x_0) = f(0). \tag{9.4}$$

For real values of ω and a we cannot extend the upper limit to infinity. We calculate the integral $K_n(a)$ in the complex z-plane. Along a contour consisting of two vertical lines, joined by the interval $0 \le z \le a$, we see that

$$K_n(a) + \int_a^{a+\mathrm{i}\infty} \mathrm{e}^{\mathrm{i}\omega z} z^{(n-1)/2}\,\mathrm{d}z = \int_0^{\mathrm{i}\infty} \mathrm{e}^{\mathrm{i}\omega z} z^{(n-1)/2}\,\mathrm{d}z \tag{9.5}$$

by means of Cauchy's theorem. In the last integral in (9.5) we put $\omega z = \mathrm{i}\lambda$ and transform it into

$$\left(\frac{\mathrm{i}}{\omega}\right)^{(n+1)/2} \left(\frac{\mathrm{i}}{\omega}\right)^{(n+1)/2} \lambda^{(n-1)/2}\,\mathrm{d}\lambda = \left(\frac{\mathrm{i}}{\omega}\right)^{(n+1)/2} \Gamma\left(\frac{n+1}{2}\right). \tag{9.6}$$

The second integral in (9.5) is transferred into

$$\left(\frac{\mathrm{i}}{\omega}\right)a^{(n-1)/2}\mathrm{e}^{\mathrm{i}\omega a}\left(\frac{\mathrm{i}}{\omega}\right)^{(n+1)/2}\left(1+\frac{\mathrm{i}\lambda}{a\omega}\right)^{(n-1)/2}\mathrm{d}\lambda$$

by the substitution of $\omega z = \omega a + \mathrm{i}\lambda$. Expanding $(1+\frac{\mathrm{i}\lambda}{a\omega})^{(n-1)/2}$ into power series in λ/ω,

$$\begin{aligned}1 &+ \frac{(n-1)}{2}\left(\frac{\mathrm{i}}{a\omega}\right)\lambda + \frac{(n-1)(n-3)}{2^2\cdot 2!}\left(\frac{\mathrm{i}}{a\omega}\right)^2\lambda^2\\ &+ \frac{(n-1)(n-3)(n-5)}{2^3\cdot 3!}\left(\frac{\mathrm{i}}{a\omega}\right)^3\lambda^3 + \cdots,\end{aligned}$$

we then obtain that

$$\begin{aligned}&\left(\frac{\mathrm{i}}{\omega}\right)a^{(n-1)/2}\mathrm{e}^{\mathrm{i}\omega a}\int_0^\infty \mathrm{e}^{-\lambda}\left(1+\frac{\mathrm{i}\lambda}{a\omega}\right)^{(n-1)/2}\mathrm{d}\lambda\\ &= \left(\frac{\mathrm{i}}{\omega}\right)a^{(n-1)/2}\mathrm{e}^{\mathrm{i}\omega a}\left\{1+\frac{(n-1)}{2}\left(\frac{\mathrm{i}}{a\omega}\right)\Gamma(2)\right.\\ &\quad\left.+\frac{(n-1)(n-3)}{2^2\cdot 2!}\left(\frac{\mathrm{i}}{a\omega}\right)^2\Gamma(3)+\cdots\right\}\\ &= \left(\frac{\mathrm{i}}{\omega}\right)a^{(n-1)/2}\mathrm{e}^{\mathrm{i}\omega a}\left\{1+\frac{(n-1)}{2}\left(\frac{\mathrm{i}}{a\omega}\right)+\frac{(n-1)(n-3)}{2^2}\left(\frac{\mathrm{i}}{a\omega}\right)^2+\cdots\right\},\end{aligned} \tag{9.7}$$

by making use of the formula $\Gamma(n+1) = \int_0^\infty \mathrm{e}^{-\lambda}\lambda^n\,\mathrm{d}\lambda = n!$ for integers n. Collecting results (9.5), (9.6), and (9.7) we have from (9.4),

$$\begin{aligned}K_n(a) &= \left(\frac{\mathrm{i}}{\omega}\right)^{(n+1)/2}\Gamma\left(\frac{n+1}{2}\right)\\ &\quad-\left(\frac{\mathrm{i}}{\omega}\right)a^{(n-1)/2}\mathrm{e}^{\mathrm{i}\omega a}\left\{1+\frac{(n-1)}{2}\left(\frac{\mathrm{i}}{a\omega}\right)+\frac{(n-1)(n-3)}{2^2}\left(\frac{\mathrm{i}}{a\omega}\right)^2+\cdots\right\}\\ &= \left(\frac{\mathrm{i}}{\omega}\right)^{(n+1)/2}\Gamma\left(\frac{n+1}{2}\right)-\mathrm{e}^{\mathrm{i}\omega a}\left(\frac{i}{\omega}\right)\\ &\quad\cdot\left\{\left(1+\left(\frac{\mathrm{i}}{\omega}\right)\frac{\mathrm{d}}{\mathrm{d}a}+\left(\frac{\mathrm{i}}{\omega}\right)^2\frac{\mathrm{d}^2}{\mathrm{d}a^2}+\cdots\right)a^{(n-1)/2}\right\}.\end{aligned} \tag{9.8}$$

From (9.2), (9.7), and (9.8), it follows that the first order term of the asymptotic expansion ($n = 0$) of J in (9.1) takes the form

$$J(0, x_0) = e^{i\omega f(0)} \frac{g(0)}{\sqrt{2f''(0)}} K_0(a) = \frac{e^{i\omega f(0)} g(0)}{\omega^{1/2}} \sqrt{\frac{\pi i}{2f''(0)}} - \frac{ie^{i\omega f(x_0)}}{\omega} \left\{ \left(1 + \left(\frac{i}{\omega}\right) \frac{d}{da} + \left(\frac{i}{\omega}\right)^2 \frac{d^2}{da^2} + \cdots \right) \frac{g(0)}{\sqrt{2f''(0)}} a^{-1/2} \right\},$$

or

$$J(0, x_0) = \frac{e^{i\omega f(0)} g(0)}{\omega^{1/2}} \sqrt{\frac{\pi i}{2f''(0)}} - \frac{ie^{i\omega f(x_0)}}{\omega} \left(g(x) \frac{dx}{dz} \right)_{z=a} + \cdots, \tag{9.9}$$

from (9.3) with α_1 being equal to zero. The first term being of order $\omega^{-1/2}$ is due to the stationary point and the following term, which has the forms resulting from the integration by parts, is due to the end point of integration.

If we extend the interval from $z = a$ to $z = \beta$, we obtain

$$e^{i\omega f(0)} \int_a^\beta e^{i\omega z} \left(g(x) \frac{dx}{dz} \right) dz$$
$$= e^{i\omega f(0)} \left[\frac{e^{i\omega z}}{i\omega} \left(g(x) \frac{dx}{dz} \right) \right]_a^\beta - \frac{e^{i\omega f(0)}}{i\omega} \int_a^\beta e^{i\omega z} \frac{d}{dz} \left(g(x) \frac{dx}{dz} \right) dz, \tag{9.10}$$

the contribution from the interval where f' is not equal to zero. Adding the two integrals from (9.9) and (9.10), we have

$$J(0, \beta) = \frac{e^{i\omega f(0)} g(0)}{\omega^{1/2}} \sqrt{\frac{\pi i}{2f''(0)}} - \frac{ie^{i\omega f(\beta)}}{\omega} \left(g(x) \frac{dx}{dz} \right)_{z=\beta} + O(1/\omega), \tag{9.11}$$

where the term of $O(1/\omega)$ is due to K_1. Hence, for large ω, the stationary point gives the largest contribution (order $\omega^{-1/2}$) while the contribution from the end points is of lower order ω^{-1}. Similarly, it is easy to see that

$$J(\alpha, 0) = \frac{e^{i\omega f(0)} g(0)}{\omega^{1/2}} \sqrt{\frac{\pi i}{2f''(0)}} + \frac{ie^{i\omega f(\alpha)}}{\omega} \left(g(x) \frac{dx}{dz} \right)_{z=\alpha} + O(1/\omega). \tag{9.12}$$

From (9.11) and (9.12), we see that if $x = \gamma$ is the only stationary point in the interval $\alpha \leq x \leq \beta$, then

$$J(\alpha, \beta) = \frac{e^{i\omega f(\gamma)} g(\gamma)}{\omega^{1/2}} \sqrt{\frac{2\pi i}{f''(\gamma)}} + O(1/\omega). \tag{9.13}$$

9.2 The Method of Characteristics

The equation

$$F(x, y, u, u_x, u_y) = 0, \tag{9.14}$$

defines a relation between the function $u = u(x, y)$ and its derivatives u_x and u_y. In order to define the function u as a solution of this equation, a set of subsidiary conditions must be prescribed. A fundamental problem for (9.14) is the Cauchy problem which consists of finding an integral surface $u(x, y)$ of (9.14) passing through a prescribed curve C in the 3-dimensional space given by

$$x = x(\lambda), \qquad y = y(\lambda), \qquad u = u(\lambda), \tag{9.15}$$

where x, y, u are smooth functions of the parameter λ.

Consider first the case that (9.14) is a quasi-linear equation, i.e.,

$$F = Au_x + Bu_y + C = 0, \tag{9.16}$$

where A, B and C are functions of x, y and u. If the Cauchy problem has a unique solution, the derivatives u_x and u_y along the curve C become functions of the parameter λ and must be determined uniquely by the conditions

$$\begin{aligned} &Au_x + Bu_y + C = 0, \quad \text{and} \\ &u_x \frac{\mathrm{d}x}{\mathrm{d}\lambda} + u_y \frac{\mathrm{d}y}{\mathrm{d}\lambda} - \frac{\mathrm{d}u}{\mathrm{d}\lambda} = 0. \end{aligned}$$

This is possible only if $\frac{1}{A}\frac{\mathrm{d}x}{\mathrm{d}\lambda} \neq \frac{1}{B}\frac{\mathrm{d}y}{\mathrm{d}\lambda}$. If however,

$$\frac{\frac{\mathrm{d}x}{\mathrm{d}\lambda}}{A} = \frac{\frac{\mathrm{d}y}{\mathrm{d}\lambda}}{B} = -\frac{\frac{\mathrm{d}u}{\mathrm{d}\lambda}}{C}, \tag{9.17}$$

the system admits a one-parameter family of solutions, i.e., the solution to the Cauchy problem (if it exists) is not unique. The curves that satisfy (9.17) are called the *characteristic* curves of the differential equation (9.16). Through a characteristic curve, several solutions of (9.16) may pass; they can, in many cases, be considered as lines of propagation of small disturbances (discontinuities in the derivatives of u). It is easily shown that the Cauchy problem for a nowhere characteristic curve C can be solved by constructing the solutions to the characteristic equations at each point of the curve and by considering the surface, $u = u(x, y)$, which is obtained in this way.

For a non-linear partial differential equation (9.14), the definition of a characteristic curve is more complicated. At a fixed point $P = (x, y, u)$ (9.14) determines a relation between u_x and u_y, i.e., it determines a one-parameter family of possible tangent planes to the integral surfaces, the solution of (9.14). The envelope of these planes form a cone with vertex at P, which is known as a *Monge* cone.

Consider again a curve C such as defined by (9.15). Then at a point on C, equation (9.14) and the compatibility condition

$$u_x \frac{dx}{d\lambda} + u_y \frac{dy}{d\lambda} = \frac{du}{d\lambda}, \tag{9.18}$$

define a number of possible tangent planes. If we assume that F is not factorable, we cannot have dependence; hence the first order derivatives u_x and u_y are always determined. The characteristic curves are obtained by considering a breakdown of the determination of the second-order derivatives. Differentiating (9.18) with respect to λ yields

$$\begin{aligned} &u_{xx}\left(\frac{dx}{d\lambda}\right)^2 + 2u_{xy}\left(\frac{dx}{d\lambda}\right)\left(\frac{dy}{d\lambda}\right) + u_{yy}\left(\frac{dy}{d\lambda}\right)^2 \\ &\quad = \frac{d^2u}{d\lambda^2} - u_x\frac{d^2x}{d\lambda^2} - u_y\frac{d^2y}{d\lambda^2}, \end{aligned} \tag{9.19}$$

and differentiating (9.14) with respect to x and y yields

$$\begin{aligned} u_{xx}F_p + u_{xy}F_q + F_u p + F_x = 0; \\ u_{xy}F_p + u_{yy}F_q + F_u q + F_y = 0, \end{aligned} \tag{9.20}$$

with $p = u_x$ and $q = u_y$. As shown before, p and q can be uniquely determined as functions of λ. Thus it follows from differentiation that

$$\begin{aligned} \frac{dp}{d\lambda} = u_{xx}\frac{dx}{d\lambda} + u_{xy}\frac{dy}{d\lambda}; \\ \frac{dq}{d\lambda} = u_{xy}\frac{dx}{d\lambda} + u_{yy}\frac{dy}{d\lambda}. \end{aligned} \tag{9.21}$$

Equations (9.20) and (9.21) give a set of four equations for the three unknowns, u_{xx}, u_{xy} and u_{yy}. The augmented matrix

$$\mathbf{A} = \begin{pmatrix} F_p & F_q & 0 & pF_u + F_x \\ 0 & F_p & F_q & qF_u + F_y \\ \frac{dx}{d\lambda} & \frac{dy}{d\lambda} & 0 & -\frac{dp}{d\lambda} \\ 0 & \frac{dx}{d\lambda} & \frac{dy}{d\lambda} & -\frac{dq}{d\lambda} \end{pmatrix}$$

will have rank generally equal to 3. If it has a rank equal to 2, then the curve C defined by (9.15) is a *characteristic*. In this case all the 3×3 submatrices of $\mathbf{A}$ are singular. This leads to the Charpit-Lagrange equations

$$\frac{dx}{F_p} = \frac{dy}{F_q} = \frac{du}{pF_p + qF_q} = \frac{-dp}{F_x + pF_u} = \frac{-dq}{F_y + qF_u}, \tag{9.22}$$

for the quantities x, y, u, p and q along the characteristic. Equations (9.22) are also known as the *characteristic* equations for the differential equation (9.14). Here again for a non-characteristic curve, the Cauchy problem is solved by constructing a surface, which is described by the characteristics issued from the points of the curve. (Note that the initial values of p and q at the point are determined!)

9.3 Singular Integral Equations

Let $\mathscr{L}$ be a segment of a closed smooth contour $\mathscr{C}$ in the complex z-plane. Let S_+ and S_- denote, respectively, the regions inside and outside the contour $\mathscr{C}$ (see Fig. 9.1). Consider the integral along $\mathscr{L}$,

$$H(z) = \frac{1}{2\pi \mathrm{i}} \int_{\mathscr{L}} \frac{h(s)}{s-z} \, \mathrm{d}s, \tag{9.23}$$

where $h(s)$ is a given Hölder continuous function (complex) along $\mathscr{L}$ such that

$$|h(s_2) - h(s_1)| \leq A|s_2 - s_1|^{\mu}, \quad \text{for } s_1, s_2 \text{ on } \mathscr{L},$$

with positive constants A and μ.

Let s_0 be a fixed point on $\mathscr{L}$, then we have for $z \in S^+$,

$$\lim_{z \to s_0} H(z) = \frac{1}{2\pi \mathrm{i}} \fint_{\mathscr{L}} \frac{h(s)}{s-s_0} \, \mathrm{d}s + \frac{1}{2\pi \mathrm{i}} \int_{\mathscr{C}^+} \frac{h(s)}{s-s_0} \, \mathrm{d}s,$$

where $\fint$ denotes the principle value, and $\mathscr{C}^+$ is a semicircle in S^- with radius ρ. The integral along $\mathscr{C}^+$ can be worked out by means of a Taylor series expansion of $h(s)$ at the point s_0. We have

$$\frac{1}{2\pi \mathrm{i}} \int_{\mathscr{C}^+} \frac{h(s)}{s-s_0} \, \mathrm{d}s = \frac{1}{2\pi \mathrm{i}} \int_{\mathscr{C}^+} \left\{ \frac{h(s_0)}{s-s_0} + h'(s_0) + \cdots \right\} \mathrm{d}s.$$

By introducing $s - s_0 = \rho \mathrm{e}^{\mathrm{i}\vartheta}$, the integral can be rewritten as

$$H^+(s_0) = \frac{1}{2} h(s_0) + \frac{1}{2\pi \mathrm{i}} \fint_{\mathscr{L}} \frac{h(s)}{s-s_0} \, \mathrm{d}s, \tag{9.24}$$

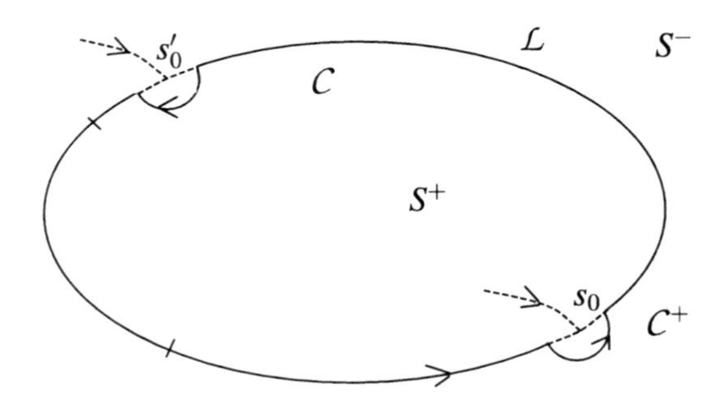

Fig. 9.1 Contour of integration

and if we approach the contour from the other side,

$$H^-(s_0) = -\frac{1}{2}h(s_0) + \frac{1}{2\pi\mathrm{i}}\int_{\mathscr{L}} \frac{h(s)}{s-s_0}\,\mathrm{d}s \tag{9.25}$$

if we let $\rho \to 0$. These are the well-known Plemelj formulas from which we see that

$$\begin{aligned} H^+(s_0) - H^-(s_0) &= h(s_0), \\ H^+(s_0) + H^-(s_0) &= \frac{1}{\pi\mathrm{i}}\int_{\mathscr{L}} \frac{h(s)}{s-s_0}\,\mathrm{d}s. \end{aligned} \tag{9.26}$$

Considering now the relation

$$H^+(s) = G(s)H^-(s) + g(s), \quad \text{on } \mathscr{L}, \tag{9.27}$$

where G and g are given functions on $\mathscr{L}$. We are interested here in the determination of $H(z)$ as an analytic continuation of the functions H_+ and H_- on $\mathscr{L}$. This is the so-called *Hilbert* problem. This can be done easily by introducing a function $\chi(z)$ which obeys the homogeneous relation

$$G(s) = \frac{\chi^+(s)}{\chi^-(s)}, \quad \text{on } \mathscr{L},$$

from which it follows that

$$\frac{H^+(s)}{\chi^+(s)} - \frac{H^-(s)}{\chi^-(s)} = \frac{g(s)}{\chi^+(s)}.$$

Thus, if

$$H(z) = \frac{\chi(z)}{2\pi\mathrm{i}}\left\{\int_{\mathscr{L}} \frac{g(s)}{\chi^+(s)(s-z)}\,\mathrm{d}s + P(z)\right\}, \tag{9.28}$$

then $H(z)$ is the corresponding analytic continuation for any arbitrary analytic function $P(z)$.

9.4 The Two-Dimensional Green's Function

Here we derive the two-dimensional version of the function of Green $\mathscr{G}(x, y; \xi, \eta)$ as used in this paper. This '*source*' function is a solution of

$$\mathscr{G}_{xx} + \mathscr{G}_{yy} = 2\pi\delta(x-\xi, y-\eta), \tag{9.29}$$

with boundary conditions

$$K\mathscr{G} - \mathscr{G}_y = 0 \quad \text{at } y = 0, \tag{9.30}$$

$$\mathscr{G}_y = 0 \quad \text{at } y = -h. \tag{9.31}$$

We introduce the Fourier transform of $\mathscr{G}$,

$$\tilde{\mathscr{G}}(y;\eta) = \frac{1}{2\pi}\int_{-\infty}^{\infty} \mathscr{G}(x,y;\xi,\eta)\mathrm{e}^{-\mathrm{i}\gamma x}\,\mathrm{d}x. \tag{9.32}$$

This transformed Green's function satisfies the conditions

$$\begin{aligned}
&\tilde{\mathscr{G}}_{yy} - \gamma^2\tilde{\mathscr{G}} = 0 \quad && \text{for } y \neq \eta,\\
&K\tilde{\mathscr{G}} - \tilde{\mathscr{G}}_y = 0 && \text{at } y = 0,\\
&\tilde{\mathscr{G}}_y = 0 && \text{at } y = -h,\\
&\lim_{\varepsilon\to 0}(\tilde{\mathscr{G}}(\eta+\varepsilon;\eta) - \tilde{\mathscr{G}}(\eta-\varepsilon;\eta)) = 0,\\
&\lim_{\varepsilon\to 0}(\tilde{\mathscr{G}}_y(\eta+\varepsilon;\eta) - \tilde{\mathscr{G}}_y(\eta-\varepsilon;\eta)) = \mathrm{e}^{-\mathrm{i}\gamma\xi}.
\end{aligned} \tag{9.33}$$

The solution of this equations is

$$\tilde{\mathscr{G}}(y;\eta) = \frac{1}{\gamma}\frac{K\sinh\gamma y + \gamma\cosh\gamma y}{K\cosh\gamma h - \gamma\sinh\gamma h}\cosh\gamma(\eta+h)\mathrm{e}^{-\mathrm{i}\gamma\xi} \quad \text{for } y > \eta, \tag{9.34}$$

$$\tilde{\mathscr{G}}(y;\eta) = \frac{1}{\gamma}\frac{K\sinh\gamma \eta + \gamma\cosh\gamma \eta}{K\cosh\gamma h - \gamma\sinh\gamma h}\cosh\gamma(y+h)\mathrm{e}^{-\mathrm{i}\gamma\xi} \quad \text{for } y < \eta. \tag{9.35}$$

Then we transform back to the x-variable. This results in (6.13) and (6.14). The contour of integration passes above or underneath the singularities on the real axis. The choice of this contour is determined by the radiation condition. For $x > \xi$ the waves travel in the positive x-direction, while for $x < \xi$ the waves travel in the negative x-direction. Therefore the contour passes the negative real pole above and the positive real pole below. Closure of the contour in the complex γ-plane leads to (6.15).

9.5 Simplification of the Set of Algebraic Equations

To obtain accurate solutions of the set of equations (6.26)–(6.28) one must get rid of the terms that consist of subtraction of large numbers. To achieve this goal we use the dispersion relation (6.10), $\gamma\tanh(\gamma h) = K$. Making use of the relation

$$\cosh(\gamma h)^2 - \sinh(\gamma h)^2 = 1$$

one obtains for the zeros $\gamma = k_i$ for $i = 0, 1, \ldots,$

$$\cosh(k_i h) = \frac{(-1)^i k_i}{\sqrt{k_i^2 - K^2}} \quad \text{and} \quad \sinh(k_i h) = \frac{(-1)^i K}{\sqrt{k_i^2 - K^2}}.$$

We also use

$$\cosh\gamma(h-d) = \cosh\gamma h\cosh\gamma d - \sinh\gamma h\sinh\gamma d,$$
$$\sinh\gamma(h-d) = \sinh\gamma h\cosh\gamma d - \cosh\gamma h\sinh\gamma d.$$

One can see that for large values depth k_0 is very close to K and the accuracy is improved if one divides out the large term analytically. The results in Fig. 6.21 can not be obtained without this simplification.

References

1. R.F. Beck, Forces and moments on a ship sailing in a shallow channel. *J. Ship Res.* **21** (1977)
2. S. Bochner, *Vorlesungen über Fouriersche Integrale* (Chelsea, New York, 1948)
3. J. Grue, E. Palme, The mean drift force and yaw moment on marine structures in waves and current. J. Fluid Mech. **250**, 121–142 (1993)
4. A.J. Hermans, A boundary element method for the interaction of free-surface waves with a very large floating flexible platform. J. Fluids Struct. **14**, 943–956 (2000)
5. A.J. Hermans, Interaction of free-surface waves with a floating dock. J. Eng. Math. **45**, 39–53 (2003)
6. A.J. Hermans, The ray method for the deflection of a floating flexible platform in short waves. J. Fluids Struct. **17**, 593–602 (2003)
7. H.W. Hoogstraten, *On non-linear dispersive water waves*. Thesis, Delft University of Technology, 1968
8. C. Hyuck, C.M. Linton, Interaction between water waves and elastic plates: Using the residue calculus technique, in *Proceedings of the 18th IWWWFB*, ed. by A.H. Clément, P. Ferrant (Ecole Centrale de Nantes, Nantes, 2003)
9. H. Maruo, The drift of a body floating in waves. J. Ship Res. **4**, 1–10 (1960)
10. C.C. Mei, J.L. Black, Scattering of surface waves by rectangular obstacles in waters of finite depth. J. Fluid Mech. **38**, 499–511 (1969)
11. M.H. Meylan, V.A. Squire, The response of a thick flexible raft to ocean waves. Int. J. Offshore Polar Eng. **5**, 198–203 (1995)
12. J.N. Newman, Lateral motion of a slender body between two parallel walls. J. Fluid Mech. **39**, 97–115 (1969)
13. J. Nossen, J. Grue, E. Palme, Wave forces on three-dimensional floating bodies with small forward speed. J. Fluid Mech. **227**, 135–160 (1991)
14. T.F. Ogilvie, Recent progress toward the understanding and prediction of ship motion, in *Fifth Symposium on Naval Hydrodynamics*, ACR-112, ONR (1964)
15. H.J. Prins, *Time-domain calculations of drift forces and moments*. PhD Thesis, TU Delft, 1995. ISBN 90-9007986-6
16. M. Roseau, *Asymptotic Wave Theory* (North-Holland, Amsterdam, 1976)
17. J.J. Stoker, *Water Waves* (Interscience, New York, 1957)
18. E.O. Tuck, Shallow water theory past slender bodies. J. Fluid Mech. **26**, 81–95 (1966)
19. J.V. Wehausen, E.V. Laitone, Surface waves, in *Encyclopedia of Physics*, vol. 9 (Springer, Berlin, 1960), pp. 446–814. Also http://www.coe.berkeley.edu/SurfaceWaves/
20. N. Wiener, *The Fourier Integral and Certain of Its Applications* (Dover, New York, 1933)

A.J. Hermans, *Water Waves and Ship Hydrodynamics*,
DOI 10.1007/978-94-007-0096-3,

21. E.T. Whittaker, G.N. Watson, *A Course of Modern Analysis*, 4th edn. (University Press, Cambridge, 1958)
22. R.W. Yeung, Application of slender body theory of ships moving in restricted shallow water, in *Proc. of Symposium on Aspects of Navigability*, Delft (1978)
23. R.W. Yeung, On the interaction of slender ships in shallow water. J. Fluid Mech. **85**, 143–159 (1978)

Index

A.J. Hermans, *Water Waves and Ship Hydrodynamics*,
DOI 10.1007/978-94-007-0096-3, © Springer Science+Business Media B.V. 2011

W